AI Longevity Ethics & Risks

The Hidden Trade-Offs, Biases, and Consequences of AI-Driven Life Extension

By Joe Correa

Published by Live Stronger Faster

This book is for educational and informational purposes only. The author and publisher are not engaged in rendering legal, financial, or other professional advice. Readers should consult a qualified professional before making any medical or financial decisions.

First Edition

Table of Contents

Introduction

The pursuit of longevity has always been a double-edged ambition. From ancient myths promising immortality to modern medicine's steady extension of human lifespan, every leap forward has carried both promise and peril. Artificial intelligence now represents the most powerful catalyst humanity has ever introduced into this pursuit. Unlike past medical advances that operated incrementally—one drug, one procedure, one protocol at a time—AI acts systemically. It learns, predicts, optimizes, and intervenes across entire biological, social, and economic systems simultaneously. This shift transforms longevity from a personal aspiration into a civilization-scale experiment, one that raises ethical questions far deeper than how long humans can live. The real question is how we *should* live longer, who benefits from that extension, and what unseen costs may follow.

AI-driven longevity systems do not merely treat disease; they shape decision-making itself. Algorithms determine which patients receive early interventions, which biomarkers are deemed significant, which lives are statistically "worth" investing in, and which populations are prioritized in research and funding. These decisions are rarely neutral. They are trained on historical data, shaped by economic incentives, and influenced by the values—explicit or implicit—of the institutions deploying them. When AI begins to influence lifespan itself, ethical considerations move beyond fairness in healthcare and into questions of justice, power, and human dignity. Longevity ceases to be a biological outcome and becomes a reflection of societal values encoded into machines.

One of the most profound risks of AI-driven life extension lies in the illusion of objectivity. Algorithms often appear impartial because they rely on data and mathematics rather than human judgment. Yet data is a record of the past, not a vision of what ought to be. Historical healthcare data reflects systemic inequality, underrepresentation, misdiagnosis, and uneven access to care. When AI systems learn from this data, they risk amplifying these disparities rather than correcting them. In the context of longevity, this could mean that life-extending technologies disproportionately benefit certain socioeconomic groups, ethnicities, or geographic regions, effectively hardcoding inequality into the future of human aging.

Beyond bias, there is the ethical dilemma of prediction itself. AI longevity systems increasingly forecast disease risk, cognitive decline, biological age acceleration, and mortality timelines. While such predictions can empower preventive care, they also raise questions about psychological burden, autonomy, and consent. Knowing one's probabilistic future is not the same as knowing a diagnosis. Predictive models can influence life choices, insurance coverage, employment decisions, and personal identity long before any disease manifests. When AI tells someone they are statistically likely to age faster or live shorter, does that information liberate them—or constrain them? And who controls how that information is used?

Another layer of ethical complexity emerges when longevity becomes programmable. AI systems are already optimizing sleep, nutrition, exercise, stress regulation, and medication adherence. As these systems grow more sophisticated, they begin to nudge human behavior continuously, subtly shaping daily decisions in the name of health optimization. At what point does guidance become coercion? When an AI system determines that certain behaviors statistically shorten lifespan, should it merely inform—or should it intervene? The line between empowerment and paternalism grows thinner as AI systems gain authority over what constitutes a "healthy" life.

The commercialization of AI longevity technologies introduces additional risks. Venture-backed startups, pharmaceutical partnerships, and data-driven health platforms operate under financial pressures that may conflict with long-term human well-being. Longevity data—genomic profiles, biometric streams, cognitive assessments—represents one of the most valuable datasets imaginable. The ethical handling of this data is not simply a matter of privacy but of power. Who owns your biological future when your aging trajectory is stored, analyzed, and monetized by private systems? And what happens when profit incentives prioritize extension of life metrics over quality of life?

AI longevity also challenges foundational ethical frameworks in medicine. Traditional medical ethics rest on principles such as beneficence, non-maleficence, autonomy, and justice. AI systems complicate each of these. Beneficence becomes probabilistic rather than certain. Harm may be delayed, diffuse, or statistical rather than immediate. Autonomy is influenced by algorithmic recommendations that shape choices invisibly. Justice must now account not only for access to care but for access to *time itself*. When lifespan becomes unevenly distributed, inequality is no longer measured only in income or opportunity, but in years of life.

There is also the existential dimension of AI-driven longevity to consider. Extending human life does not merely add years; it reshapes narratives of meaning, purpose, and legacy. Societies built around predictable life stages—education, work, retirement, aging—must renegotiate their structures when those stages stretch or blur. Ethical risk emerges when technological capability outpaces cultural adaptation. Without thoughtful governance, longer lives could intensify burnout, economic stagnation, generational tension, or psychological fatigue rather than flourishing.

Importantly, ethical risk does not imply ethical failure. AI-driven longevity holds extraordinary potential to reduce suffering, prevent disease, and enhance human potential across the lifespan. But realizing that potential requires intentional design, transparent governance, and continuous ethical reflection. This book does not argue against AI longevity; it argues against unexamined AI longevity. Progress without ethics is not advancement—it is acceleration without direction.

Throughout this book, we will explore the hidden trade-offs embedded in AI-driven life extension systems. We will examine bias, surveillance, consent, data ownership, algorithmic authority, and the moral implications of extending life in an unequal world. We will analyze not only what AI *can* do for longevity, but what it *should* do—and where restraint, regulation, or redesign may be necessary. The goal is not to slow innovation, but to guide it toward outcomes that honor human dignity, equity, and long-term societal health.

As AI continues to redefine aging, ethics becomes the true limiting factor—not technology. The future of longevity will not be decided solely by algorithms, models, or breakthroughs, but by the values we choose to encode into them. This book invites readers to step into that responsibility, to question assumptions, and to engage critically with a future in which living longer is no longer a biological accident, but a technological choice.

Chapter 1

The Ethics of Who Lives Longer

The moment humanity gained the ability to influence lifespan, ethics became inseparable from longevity. For most of history, how long a person lived depended largely on chance: genetics, environment, nutrition, and the randomness of disease. Medicine gradually shifted this reality by preventing infections, repairing injuries, and treating chronic conditions. Yet these advances still operated within the natural variability of human biology. Artificial intelligence changes this equation fundamentally. AI does not merely treat illness; it stratifies risk, allocates intervention, predicts decline, and optimizes health trajectories across populations. In doing so, it introduces a new ethical question at the center of longevity science: when technology can extend life selectively, who lives longer—and why?

AI-driven longevity systems function by identifying patterns in biological and behavioral data that correlate with health outcomes. These patterns allow algorithms to recommend interventions, prioritize patients, and allocate resources with unprecedented precision. On the surface, this appears ethically sound. Preventive care can be targeted to those most at risk, treatments can be delivered earlier, and healthspan can be improved efficiently. However, the ethical tension arises from the fact that prediction inherently creates hierarchy. Individuals are categorized by risk scores, projected lifespan, biological age acceleration, or intervention responsiveness. These categories influence who receives attention, funding, and technological benefit. Longevity, once a shared human aspiration, becomes stratified by algorithmic assessment.

The ethical implications deepen when considering the data that trains AI longevity systems. Health datasets are not neutral reflections of biology; they are shaped by access to care, socioeconomic conditions, geographic disparities, and historical bias. Populations with greater medical access generate richer data, enabling more accurate AI models for those groups. Conversely, underrepresented populations produce sparse or incomplete datasets, reducing predictive accuracy and limiting algorithmic confidence. When AI allocates life-extending interventions based on model certainty, it may inadvertently favor those already privileged by healthcare access. Thus, the ethical risk is not overt discrimination but structural reinforcement of existing inequality—longevity advantages accumulating where data is strongest.

Another dimension of this ethical challenge lies in economic valuation. Longevity technologies are costly to develop and deploy, involving advanced diagnostics, continuous monitoring systems, and personalized interventions. Market forces inevitably shape their distribution. Insurance systems, healthcare providers, and private longevity platforms must decide which interventions are financially justified. AI models increasingly inform these decisions by estimating treatment effectiveness, cost-benefit ratios, and projected years of life gained. When lifespan becomes quantifiable in economic terms, ethical tension emerges between efficiency and equity. Should resources prioritize those with the highest probability of benefit, or those with the greatest need? AI can calculate both, but it cannot resolve the moral choice between them.

The concept of fairness itself becomes more complex in AI longevity contexts. Traditional healthcare ethics often emphasize equal treatment for equal need. AI systems, however, operate on probabilistic optimization rather than equal distribution. They recommend interventions where outcomes are most likely to improve measurable health metrics. This may lead to scenarios in which individuals with lower predicted responsiveness receive fewer resources. While statistically rational, such decisions raise ethical concerns about abandonment or reduced support for those already biologically disadvantaged. Longevity ethics must therefore grapple with whether fairness means maximizing total life-years or ensuring equitable opportunity for extension across all individuals.

Predictive longevity analytics introduce further ethical layers. AI can estimate biological age acceleration, future disease probability, and mortality risk decades in advance. Such forecasts influence behavior, identity, and social perception. Individuals labeled as high-risk may face psychological burden, altered life planning, or stigmatization. Employers, insurers, or institutions could misuse predictive longevity data to assess long-term productivity or financial liability. Even when predictions are probabilistic rather than deterministic, their perceived authority can shape societal attitudes toward individuals' expected lifespans. Ethical frameworks must therefore address not only the accuracy of prediction but its social consequences.

The authority granted to AI systems in longevity decision-making represents another ethical frontier. As models outperform human clinicians in pattern detection, their recommendations may gain institutional trust and operational priority. Clinical judgment may gradually defer to algorithmic assessment. Yet AI does not possess moral reasoning; it optimizes according to objectives embedded in its design. If those objectives prioritize measurable longevity metrics—years of life gained, disease-free survival, or biological age reduction—they may overlook qualitative dimensions such as autonomy, meaning, or subjective well-being. Ethical risk arises when algorithmic authority narrows the definition of a good life to quantifiable health outcomes.

Consent in AI longevity contexts is uniquely complex. Individuals may consent to data collection or intervention without fully understanding long-term implications. Longevity systems often integrate multiple data streams—genomic, metabolic, behavioral, environmental—creating continuously evolving health profiles. These profiles can generate new insights long after initial consent. Furthermore, predictive longevity analytics may affect relatives who share genetic or environmental characteristics. Consent therefore extends beyond individuals to families and potentially communities. Ethical governance must consider whether consent can remain valid when data use and predictive capability expand over time.

The ethics of responsibility also shift in AI-optimized longevity. When systems provide continuous health guidance—dietary recommendations, activity targets, sleep optimization—individuals are increasingly aware of behaviors influencing lifespan. This awareness can empower but also burden. Health becomes framed as a personal optimization project, with deviation from algorithmic recommendations interpreted as self-harm or negligence. Societal narratives may evolve toward moralizing longevity outcomes, attributing shorter lifespans to individual failure rather than structural factors. Ethical reflection must therefore guard against transforming longevity from a shared societal goal into a measure of personal virtue.

Global disparities intensify these ethical considerations. AI longevity technologies emerge primarily in wealthier regions with advanced healthcare infrastructure and data ecosystems. Their benefits may reach lower-resource settings slowly or unevenly. If lifespan extension accelerates in technologically advanced populations while others remain constrained by existing health challenges, global longevity gaps could widen significantly. Such divergence raises profound ethical questions about justice across nations and generations. The extension of human life in one region should not implicitly diminish the value or urgency of life elsewhere.

Ultimately, the ethics of who lives longer cannot be resolved solely through technical design. It requires deliberate societal choice. AI longevity systems reflect priorities embedded by their creators, funders, regulators, and users. Decisions about data inclusion, intervention thresholds, accessibility, and resource allocation shape lifespan outcomes. Ethical longevity therefore depends on transparent governance frameworks, inclusive datasets, equitable distribution strategies, and continuous oversight. The challenge is not merely preventing harm but ensuring that extended life remains a collective human advancement rather than a selectively distributed privilege. This chapter establishes a foundational principle for AI longevity ethics: technological capacity to extend life must be accompanied by ethical commitment to equitable extension. The question is not whether AI can help humans live longer—it already does. The question is whether humanity chooses to distribute that capacity fairly, responsibly, and with awareness of its broader consequences. The future of longevity will be defined not only by innovation, but by how societies decide whose years are extended and whose are not.

The Years That Were Counted

In the year 2042, lifespan forecasts were as common as weather predictions. Every citizen carried a longevity profile generated by the Global Health Grid, an AI system integrating genomic sequencing, lifelong biometric monitoring, environmental exposure mapping, and behavioral analytics. At birth, each person received an evolving projection of biological age trajectory and expected lifespan range. Governments claimed the system democratized preventive care. Critics argued it quietly ranked human lives.

Elena Varga first saw her longevity forecast at sixteen. The school nurse summoned her after a routine health scan flagged accelerated biological aging markers. The AI interface displayed a simple statement: projected lifespan percentile—38th. Risk cluster—metabolic and inflammatory. Intervention priority—moderate. Elena stared at the numbers without fully understanding them, but she recognized the implication. Most of her classmates were ranked above average. She was not.

Her family dismissed the report at first. Elena was athletic, curious, and resilient. She rarely fell ill. Yet over time the projection followed her invisibly. University health services recommended preventive programs she could not afford. Insurance plans adjusted premiums based on longevity risk tiers. Employers offering long-horizon training programs quietly favored applicants with higher projected lifespan percentiles. No law mandated such practices; they emerged organically from a culture that trusted the Grid's forecasts.

At twenty-seven, Elena worked as a materials engineer designing bioadaptive housing surfaces. Her performance reviews were excellent, yet promotions passed her by. Colleagues with similar experience advanced faster. A friend in human resources confided the truth over coffee: leadership investment models incorporated longevity stability metrics.

Employees expected to remain healthy longer justified higher long-term training investment. Elena's projected trajectory reduced her perceived return on development.

The realization did not break her immediately. Instead, it settled into her thinking as a subtle boundary. She avoided career paths requiring decades of accumulation. She postponed long-term commitments. Relationships felt shadowed by probability curves she could not escape. She told herself she did not believe the forecast, yet she lived around it.

One autumn evening, Elena attended a public lecture by Dr. Ishan Rao, an AI ethicist studying longevity allocation systems. He spoke about statistical life valuation, algorithmic stratification, and the risk of self-fulfilling prophecies. "Prediction," he said, "becomes destiny when societies reorganize around it." Elena felt as if someone had articulated the silent architecture shaping her life. After the talk, she waited in line and showed him her longevity profile. He studied it quietly, then asked, "Has anyone explained the uncertainty range to you?"

No one had. The percentile appeared authoritative, precise. Dr. Rao described the confidence intervals, data sparsity in certain populations, and bias amplification in early models. Elena's genetic ancestry, he explained, was underrepresented in historical datasets used to train metabolic aging predictors. Her forecast carried wider uncertainty than the interface conveyed. She was not necessarily aging faster; the system simply knew less about people like her.

The revelation destabilized her assumptions. For years she had interpreted the forecast as biological truth. Now she saw it as probabilistic inference shaped by incomplete data. The number that had silently constrained her choices was less certain than it appeared. Dr. Rao invited her to participate in a research initiative examining algorithmic longevity bias. She agreed, partly out of curiosity, partly from the faint stirrings of reclaiming authorship over her own timeline.

The project analyzed thousands of profiles across demographic groups, comparing predicted and actual biological aging markers over time. Patterns emerged: underrepresented populations systematically received lower longevity projections with higher error margins. The AI models, optimized for predictive accuracy on dominant datasets, generalized poorly to others. Elena's case was not anomalous; it was structural. The Grid's forecasts had quietly encoded data inequality into lifespan expectations.

Publishing these findings sparked global debate. Governments defended the Grid's benefits in preventive care. Advocates demanded transparency and recalibration. Longevity percentile displays were redesigned to include uncertainty visualization and demographic representation indicators. Investment algorithms incorporating lifespan forecasts faced regulatory scrutiny. For the first time, societies confronted the ethical weight of predictive longevity systems. For Elena, the impact was both external and internal. Career barriers eased as policy changes limited longevity-based investment metrics. More importantly, the psychological boundary dissolved. She enrolled in a doctoral program studying adaptive biomaterials for regenerative environments—work she once considered too long-term for her projected lifespan. Her research focused on designing living surfaces that supported microbiome diversity and metabolic resilience in urban populations historically underrepresented in health datasets. She was, in essence, extending the biological context the Grid had failed to understand.

Years passed. Elena's biological markers stabilized, then improved. Whether due to lifestyle, environment, or statistical regression, her updated longevity profile rose into average percentiles. The shift mattered less than she expected. She no longer organized her life around the number. The forecast had become a tool rather than a verdict.

At forty-nine, Elena stood on the stage of the International Longevity Ethics Forum beside Dr. Rao, presenting the Adaptive Data Equity Framework now adopted by multiple health systems. It required continuous dataset diversification, bias auditing, and probabilistic transparency in lifespan predictions. "Longevity technology," she told the audience, "must not decide who deserves time. It must expand time responsibly for everyone." After the session, a young attendee approached her with trembling hands, holding a newly issued longevity profile marked below average. Elena recognized the look—the quiet collapse of imagined years. She smiled gently and said the words she wished someone had told her decades earlier: "This number describes what the system currently understands, not what your life will be. You are larger than its data."

That evening Elena walked along the river outside the conference center. The city lights reflected across the water in shifting patterns, never fixed, always reforming. She thought about the years once counted against her, the paths she almost did not take, the future she nearly surrendered to probability. Longevity, she realized, was not merely the extension of biological time. It was the freedom to inhabit time without algorithmic ceilings.

The Grid still existed, more accurate and more ethical than before. Forecasts guided preventive care across the world. Yet Elena knew the deeper lesson remained unfinished. Every generation would inherit predictive systems shaping expectations of lifespan and worth. The ethical task would never end: ensuring that technology measuring life never diminished its possibility. She paused at the water's edge, watching reflections lengthen into darkness. The future stretched outward, unranked and unbounded. And for the first time since adolescence, Elena felt not predicted years ahead of her—but open years.

Chapter 2

The Price of Extended Life

The extension of human lifespan has never been free. Every increase in longevity across history has required resources—nutrition, sanitation, medicine, infrastructure, and social stability. Artificial intelligence introduces a new layer to this equation by accelerating the development and personalization of life-extending technologies. AI-driven longevity systems require advanced diagnostics, continuous data monitoring, computational modeling, and targeted interventions that operate across decades of an individual's life. As these systems mature, they shift longevity from a natural outcome of collective health progress into a technologically mediated service. This transformation raises a central ethical question: when longer life becomes technologically purchasable or economically mediated, what is the true price of extended life, and who is expected to pay it?

AI longevity systems operate through a combination of predictive analytics and continuous optimization. Individuals generate vast streams of biological and behavioral data—genomic sequencing, metabolic markers, microbiome composition, activity patterns, sleep architecture, environmental exposure, and cognitive performance metrics.

AI models analyze these inputs to forecast disease risk and recommend interventions long before symptoms emerge. While this preventive model promises unprecedented healthspan gains, it also establishes a persistent dependency on monitoring technologies and clinical infrastructure. Longevity becomes less an episodic medical event and more an ongoing subscription to health optimization. Ethical concern arises when access to extended life increasingly correlates with access to such systems.

Economic stratification is therefore one of the most significant risks in AI-driven longevity. High-resource populations adopt advanced monitoring devices, personalized therapeutics, regenerative interventions, and AI-guided lifestyle optimization earlier and more comprehensively. Lower-resource populations may receive only partial or delayed access. Even when technologies eventually diffuse, early adopters accumulate compounding advantages. Preventive interventions applied decades earlier alter biological trajectories permanently. The result is not merely a temporary health gap but a divergence in lifetime health capital. Ethical frameworks must confront whether AI longevity amplifies existing socioeconomic gradients in lifespan.

The financial architecture of longevity technologies further complicates distribution. Many AI-driven interventions emerge through private investment, pharmaceutical partnerships, and proprietary data platforms. Intellectual property protections and market incentives shape pricing structures that prioritize return on investment. This dynamic is not inherently unethical; innovation often depends on economic viability. However, when applied to lifespan itself, profit-driven distribution raises moral tension. If life-extending interventions remain expensive for extended periods, longevity becomes partially commodified. Societies must then confront whether years of life should be subject to market allocation in the same manner as consumer goods or elective services.

Insurance systems face unprecedented challenges in this context. Traditional health insurance distributes risk across populations based on probabilistic illness occurrence. AI longevity technologies invert this model by predicting individual risk and enabling targeted prevention. Insurers may seek to incentivize participation in longevity monitoring programs, adjust premiums based on adherence to optimization protocols, or limit coverage for individuals declining recommended interventions. Such practices could create pressure toward algorithmically guided lifestyles. Ethical risk arises when financial structures effectively coerce individuals into continuous surveillance and behavioral conformity in exchange for access to healthspan extension. The concept of personal responsibility in health also evolves under AI longevity. When predictive systems identify behaviors correlated with accelerated aging, individuals may be expected to modify those behaviors to maintain coverage or social approval. This expectation assumes equal capacity to comply with recommendations, ignoring disparities in environment, occupation, culture, and resources. A person working multiple jobs with limited access to nutritious food or safe exercise environments cannot easily implement optimization protocols designed around ideal conditions. Ethical evaluation must therefore consider whether AI longevity frameworks inadvertently penalize those constrained by structural factors beyond individual control.

Another dimension of the price of extended life lies in data extraction. Longevity systems require extensive personal information across decades. Biological samples, digital biomarkers, behavioral records, and environmental exposure histories become part of integrated datasets used to refine AI models. Individuals contribute this data both as patients and as research participants, often without direct compensation or ownership rights. Corporations and institutions may derive significant value from aggregated longevity data while individuals receive primarily indirect benefits. Ethical questions emerge regarding data ownership, consent longevity, and the fairness of value distribution derived from human biological information.

There is also a temporal dimension to longevity cost. AI-driven interventions may extend lifespan significantly, but longer life entails extended healthcare needs, social support, and economic participation. Societies must sustain aging populations over longer periods, potentially reshaping retirement systems, labor markets, and intergenerational resource allocation. If longevity gains concentrate among certain groups, demographic imbalances may intensify. Ethical analysis must therefore consider not only individual access to life extension but societal capacity to integrate extended life fairly across generations.

The psychological cost of longevity optimization should not be overlooked. Continuous monitoring and intervention create a mindset in which health becomes a perpetual project requiring vigilance and compliance. Individuals may feel responsible for maintaining optimal biomarker ranges and preventing predicted decline. Deviations from recommended protocols may evoke guilt or anxiety. Life lived under constant optimization pressure risks transforming well-being into performance metrics. Ethical frameworks must ask whether extended lifespan achieved through perpetual self-surveillance constitutes genuine flourishing or a technologically sustained burden. Cultural perspectives on aging further influence perceptions of longevity value. In some traditions, aging is associated with wisdom, community role transition, and acceptance of mortality. AI longevity paradigms often frame aging primarily as a modifiable risk state. This reframing may marginalize cultural meanings of later life that emphasize relational or spiritual dimensions rather than biological optimization. Ethical integration of longevity technologies requires respect for diverse conceptions of a good life across cultures, rather than imposing a singular technologically oriented model of aging.

Global equity remains a critical concern. The development of AI longevity technologies relies heavily on data and infrastructure from wealthier nations. If resulting interventions remain concentrated in those regions, global lifespan disparities may widen. Ethical responsibility extends beyond national borders, particularly when knowledge derived from global populations contributes to innovation. Mechanisms for international access, technology transfer, and equitable deployment must be considered integral to ethical longevity advancement.

Ultimately, the price of extended life cannot be measured solely in financial terms. It encompasses economic resources, personal data, behavioral autonomy, psychological burden, and societal adaptation. AI longevity technologies promise extraordinary gains in healthspan and lifespan, yet without deliberate ethical design they risk transforming life extension into a selectively distributed privilege requiring continuous economic and behavioral investment. The challenge for societies is to ensure that extended life remains a shared human good rather than an exclusive outcome of technological purchasing power.

This chapter establishes a second foundational principle of AI longevity ethics: the extension of human life should not depend primarily on wealth, compliance capacity, or data capital. Longevity technologies must be guided by distribution models that balance innovation incentives with equitable access. The measure of ethical success in AI longevity will not be how far life can be extended, but how broadly and fairly that extension is made possible.

The Years That Could Be Bought

By 2051, longevity subscriptions had replaced health insurance across most developed regions. Citizens enrolled in tiered programs offering varying degrees of biological monitoring, predictive modeling, and regenerative intervention coverage. The tiers were described in neutral language—Essential, Adaptive, Horizon—but everyone understood their implications. Essential preserved baseline health. Adaptive optimized aging trajectories. Horizon actively extended lifespan through experimental and advanced therapies. No law required enrollment, yet opting out meant accepting unoptimized aging in a world where extended life had become expected.

Mateo Alvarez lived in the Essential tier.

He worked as a transit systems technician maintaining autonomous rail corridors beneath the city. The job was stable, physically demanding, and poorly paid relative to the cognitive industries above ground. Essential coverage provided annual biomarker screening, standard preventive protocols, and limited AI health guidance. Mateo wore a basic metabolic band issued by the program. It tracked sleep and activity but lacked the advanced genomic integration and continuous cellular monitoring available in higher tiers. He rarely thought about it until his older sister Sofia invited him to her Horizon renewal consultation.

Sofia was a neuroarchitect designing cognitive environments for immersive learning networks. Horizon coverage had stabilized her biological age for nearly a decade. Her skin remained unlined, her cognition sharp, her energy unflagging. The clinic resembled a quiet gallery of light and glass. AI projections displayed her cellular repair indices and projected lifespan horizon: 112 years. Mateo's last Essential forecast had estimated eighty-three. During the consultation, Sofia's physician mentioned Mateo's presence and asked casually whether he had considered upgrading tiers. Mateo laughed, deflecting. The physician did not press, yet the suggestion lingered. On the journey home through the maintenance tunnels, Mateo became acutely aware of physical fatigue he had long ignored—the ache in his joints after climbing service ladders, the slower recovery after night shifts, the quiet heaviness in his muscles. For the first time, he wondered whether aging itself had become stratified.

Weeks later, a notice arrived: Essential tier participants in his district would lose access to certain regenerative therapies due to funding reallocation toward Adaptive expansion. The therapies were modest—joint cartilage preservation, metabolic recalibration—but their removal implied a narrowing trajectory. Mateo visited the public clinic to review his profile. The AI interface projected his biological age curve gradually diverging from Adaptive and Horizon cohorts by midlife. The display used gentle colors and neutral phrasing, yet its message was unmistakable: without intervention, his aging would accelerate relative to those with higher-tier optimization.

The revelation unsettled him more than Sofia's polished longevity. It was not envy but recognition of structural difference. Longevity had become a gradient of access. Mateo returned home through streets layered with augmented advertisements offering financing plans for Adaptive upgrades. Payments spread across decades. Future labor earnings collateralized against extended healthspan. The logic was circular: live longer to work longer to pay for living longer.

He began researching alternatives. Community forums mentioned cooperative longevity collectives—groups pooling resources to access shared AI optimization tools. Most were experimental or unstable. One, however, persisted: the Tiempo Común Cooperative operating from repurposed transit depots across several cities. Their model rejected tiered subscriptions, emphasizing open-source AI longevity systems, shared diagnostics, and collective funding of interventions prioritized by need rather than income.

Mateo attended a meeting in an abandoned maintenance hall transformed into a clinic. The environment lacked Horizon's elegance but radiated purposeful care. Participants ranged from manual laborers to educators, caregivers, and displaced professionals. The cooperative's AI platform projected anonymized group health metrics and intervention priorities. Resources were allocated to members with the greatest biological age acceleration regardless of contribution level. The principle was simple: longevity as a shared commons rather than a purchased service. Initial skepticism gave way to involvement. Mateo contributed technical skills repairing monitoring hardware and maintaining sensor networks. In exchange, he received enhanced diagnostics and personalized metabolic protocols derived from the cooperative's evolving dataset. The interventions were modest compared to Horizon therapies yet precisely targeted. Over months his recovery improved, inflammation markers fell, and fatigue lessened. The cooperative's AI forecast revised his lifespan projection upward—not dramatically, but meaningfully.

The deeper transformation, however, was cultural. Cooperative members discussed aging without tier language. Lifespan was framed not as individual optimization but collective resilience. When a member required costly regenerative therapy, resources were reallocated temporarily across the group. Decisions were debated openly, balancing urgency and sustainability. Mateo witnessed an alternative ethics of longevity emerging organically: extended life pursued through mutual support rather than competitive acquisition.

News of Tiempo Común spread. Critics labeled it inefficient and technologically inferior. Yet independent studies revealed surprising outcomes: cooperative participants achieved healthspan improvements comparable to mid-tier Adaptive programs despite lower per-capita expenditure. The advantage lay in distribution logic. Interventions reached those with highest need earliest, altering trajectories before decline compounded. Longevity gains, while moderate individually, aggregated across the population.

Years passed. Mateo aged into his sixties with mobility and vitality exceeding his Essential forecast decades earlier. Sofia, still within Horizon, visited the cooperative during a research exchange. She observed group decision councils allocating regenerative access based on biological urgency. The absence of tier hierarchy unsettled her assumptions. Horizon had extended her life significantly, yet it had also isolated her trajectory from others. The cooperative extended life more modestly but bound trajectories together.

One evening Sofia and Mateo walked above ground along the elevated rail her brother maintained. City lights shimmered against the river. Sofia spoke quietly: “I always thought longevity was something you secured. Protected. Purchased.” Mateo considered the statement. “Maybe it’s something we build,” he replied. “Like infrastructure. Shared so it works for everyone.”

Months later Horizon insurers introduced cooperative-inspired distribution pilots after demographic modeling showed unsustainable divergence between tiers. Policy shifted gradually toward hybrid systems integrating market innovation with communal allocation principles. The gradient of purchasable years softened, though never disappeared entirely. Yet the ethical axis had moved: longevity was no longer viewed solely as a commodity but increasingly as a social good requiring stewardship. At seventy-eight, Mateo's latest projection exceeded ninety-five. The number mattered less than the path that produced it. He had not escaped the price of extended life; he had transformed its currency. Instead of wealth alone, the cost had included participation, contribution, and mutual care. Longevity, he realized, was not merely additional years granted by technology. It was time made possible by relationships structured around shared survival.

Standing again on the rail overlooking the city he had maintained for decades, Mateo watched trains glide beneath fading twilight. Each carriage carried lives moving forward through time, some longer, some shorter, all intertwined. He understood then that the future of longevity would not be decided solely in laboratories or markets. It would be decided wherever humans chose whether years belonged to individuals—or to humanity together.

Chapter 3

Surveillance, Autonomy, and the Quantified Life

Artificial intelligence–driven longevity systems depend fundamentally on data. The ability to predict disease, optimize health trajectories, and personalize interventions arises from continuous streams of biological and behavioral information. Wearable sensors monitor physiology, digital environments record activity patterns, genomic and metabolic profiles reveal internal states, and environmental tracking maps exposures across time. Together, these data form an integrated portrait of human life unprecedented in scope and continuity. Longevity optimization thus becomes inseparable from surveillance. The ethical question is not whether monitoring improves health—it demonstrably can—but whether a life continuously observed in the name of longevity remains fully autonomous.

Surveillance in longevity contexts differs from traditional medical observation. Historically, medical data were episodic and localized, gathered during illness or clinical visits. AI longevity systems require persistent monitoring across daily life, capturing information even in the absence of disease. This shift transforms health data from discrete medical records into ongoing behavioral archives. Sleep duration, nutritional intake, emotional variability, movement patterns, cognitive responses, and social interaction rhythms all contribute to predictive models. As monitoring expands, the boundary between healthcare and personal life dissolves. Ethical tension arises when optimization demands visibility into domains once considered private or self-directed.

The justification for such surveillance rests on predictive benefit. Early deviations in biomarker trends can signal future pathology long before symptoms arise. AI systems detect subtle correlations inaccessible to human perception. Continuous data allow interventions that prevent disease entirely rather than treating it after onset. From a purely health perspective, greater data yields greater predictive power. Yet ethical evaluation must consider proportionality. Not all health gains justify equivalent intrusion. Societies must determine how much observation is acceptable in exchange for how much extension of life or healthspan.

Autonomy is central to this balance. A core principle of ethical medicine is respect for individuals' capacity to make informed decisions about their bodies and lives. Continuous monitoring can both support and undermine this principle. On one hand, AI guidance empowers individuals with personalized insight, enabling proactive health choices. On the other, algorithmic recommendations may gradually shape behavior through nudges, alerts, and risk scores that influence perception of acceptable choices. When deviation from optimization protocols is framed as self-harm or health negligence, autonomy narrows. Individuals may feel compelled to conform to algorithmic norms even when they conflict with personal values or preferences.

The authority of predictive analytics intensifies this dynamic. Longevity AI does not merely suggest general health advice; it provides probabilistic forecasts tied to measurable outcomes. A recommendation to reduce certain foods or activities carries implicit justification: predicted reduction in lifespan or disease risk. While technically probabilistic, such forecasts often acquire perceived determinism. Individuals may internalize them as inevitable outcomes rather than modifiable probabilities. Ethical risk arises when algorithmic projections redefine personal identity and life choices around statistical expectation rather than lived intention.

Another dimension of surveillance ethics concerns data ownership and control. Longevity monitoring generates vast personal datasets valuable for research, commercial development, and institutional decision-making. Individuals may consent to data collection for health optimization without fully understanding downstream uses. Aggregated datasets refine predictive models and create economic value for corporations or healthcare systems. Yet individuals rarely retain meaningful ownership or compensation for their biological information. The ethical question extends beyond privacy toward equity: who benefits from data derived from human lives, and how are those benefits distributed?

Data permanence also complicates consent. Longevity profiles accumulate across decades, integrating historical data into evolving predictions. Information provided in youth may influence assessments in later life contexts unforeseen at the time of collection. Withdrawal of consent becomes technically difficult when data are embedded in large-scale models. Ethical governance must address whether lifelong surveillance can ever be fully voluntary when participation creates long-term informational entanglement beyond individual control.

The psychological impact of quantified life monitoring represents another ethical frontier. Continuous metrics encourage individuals to interpret their bodies through numerical indicators: heart rate variability, metabolic efficiency, cognitive latency, inflammation indices, biological age. These metrics can motivate health behavior, yet they also risk narrowing self-perception. Human experience encompasses qualitative dimensions—joy, meaning, creativity, relational connection—that resist quantification. When longevity systems emphasize measurable health optimization, individuals may prioritize metric improvement over experiential richness. Ethical longevity must therefore consider whether quantified living enhances or diminishes holistic well-being.

Social implications extend beyond individuals. Institutions may seek access to longevity monitoring data to manage risk or productivity. Employers could favor candidates with stable health trajectories. Insurers might adjust coverage based on compliance with optimization protocols. Governments could incentivize participation in surveillance-based prevention programs to reduce healthcare costs. While each application may appear rational individually, collectively they risk normalizing continuous observation as a condition of social participation. Autonomy erodes not through coercion but through systemic expectation that responsible citizens monitor and optimize themselves.

Cultural diversity complicates universal adoption of quantified longevity paradigms. Not all societies value constant self-measurement or predictive health framing. Some traditions emphasize acceptance of bodily variability and mortality as integral to human experience. AI longevity systems developed within technologically oriented cultures may implicitly encode assumptions about desirable behavior, diet, activity, or emotional regulation. Exporting such systems globally without cultural adaptation risks imposing uniform norms under the guise of health optimization. Ethical deployment requires sensitivity to pluralistic conceptions of well-being.

Importantly, surveillance in longevity contexts can be designed along a spectrum. Systems may prioritize minimal necessary data, decentralized storage, and user-controlled access. Alternatively, they may aggregate maximal data centrally for predictive accuracy. Ethical frameworks must determine acceptable trade-offs between model performance and personal autonomy. Transparency in algorithmic function, opt-out pathways, and meaningful consent mechanisms are essential to maintaining trust and agency. The goal is not eliminating monitoring—impossible in effective longevity AI—but shaping it to respect human dignity.

This chapter establishes a third foundational principle of AI longevity ethics: the extension of life should not require surrender of autonomy. Surveillance-based optimization must remain proportional, consensual, and transparent. Individuals should retain the freedom to define meaningful lives even when those lives deviate from algorithmic ideals of health efficiency. Longevity gained at the cost of self-determination risks becoming a hollow victory. The ethical future of AI longevity lies not only in how long humans live, but in how freely they live within that extended time.

The Life That Was Measured

When the Quantified Life Act passed in 2047, participation in national longevity monitoring programs became nearly universal. Citizens received integrated bio-sensor implants at adulthood linking physiological data to the public health AI. The system promised unprecedented disease prevention and lifespan extension. Compliance was voluntary in principle, yet access to many services—from employment certifications to travel insurance—gradually required active monitoring status. Most people accepted the implants with little resistance. Health gains were visible and immediate. Illness incidence fell. Lifespans lengthened. The world grew statistically healthier.

Mara Ionescu resisted.

She was a composer whose work depended on unpredictable rhythms of creation. Her days oscillated between intense focus and wandering reflection. Sleep patterns varied. Meals were irregular. Long walks without devices were her most fertile creative periods. When the implant invitation arrived, she declined politely. At first this choice carried minimal consequence. She submitted periodic clinical scans to maintain baseline health clearance. Friends teased her gently as an analog soul in a quantified era. Over time the gap widened. Opportunities requiring continuous monitoring excluded her. Performance venues partnered with longevity-certified artists under contracts guaranteeing sustained cognitive acuity projections. Insurance premiums for unmonitored individuals rose sharply. Even social invitations subtly shifted toward optimized cohorts whose schedules aligned with predictive health patterns. Mara continued composing, yet she sensed her world narrowing around an unspoken boundary: participation equaled responsibility; refusal implied risk.

Her younger brother Luca embodied the opposite path. A systems engineer, he embraced full-spectrum monitoring. His implant interfaced with environmental sensors, cognitive trainers, and metabolic regulators. Luca's biological age remained nearly static through his thirties. He encouraged Mara to reconsider. "It's not control," he insisted. "It's information. Why choose uncertainty?" She struggled to articulate her unease. The implant itself was benign. The data secure. Yet she felt that continuous measurement would reshape her relationship with time. Creativity for her required spaces unobserved even by herself.

One winter she received an invitation impossible to ignore: a commission from the National Symphony celebrating the Quantified Era. The contract required implant certification. Accepting meant joining the monitored population. Declining meant forfeiting the largest platform of her career. She walked the city for days weighing the choice. Surveillance promised stability and opportunity. Refusal preserved autonomy but risked marginalization. The dilemma was not technological but existential: could a life devoted to unstructured experience coexist with a society optimized around measurement?

On the final evening before her decision deadline, Mara visited Luca's apartment. His living space hummed softly with bioadaptive systems adjusting light, air composition, and sound frequencies based on his physiological state. He seemed effortlessly well. She envied the ease but sensed its architecture: a life aligned continuously with predictive guidance. Luca prepared tea while his implant projected hydration and metabolic indicators into peripheral vision. He spoke gently: "The implant won't change who you are." Mara wondered whether any constant mirror left identity untouched.

She returned home and opened her notebooks. Pages filled with sketches of melodies born during sleepless nights, wandering walks, unplanned encounters. None had been optimized. All had emerged from states unpredictable and often physiologically inefficient. She realized her fear was not surveillance itself but the subtle redefinition of value. In a measured world, health metrics acquired authority over time use. Activities lacking optimization justification—aimless reflection, irregular rhythm, emotional immersion—risked classification as deviation. Her music lived precisely in those spaces.

The next morning she declined the implant and the commission.

Consequences followed predictably. Institutional venues closed. Funding dwindled. Yet unexpected paths opened. Independent collectives of unmonitored creators, philosophers, and experiential practitioners formed networks outside optimization culture. They called themselves the Unmeasured Guild. Members believed human flourishing required domains beyond predictive framing. They gathered in remote retreats where devices were absent and time unstructured. Mara's compositions evolved there, shaped by silence and shared presence rather than biometrics.

Years passed. The quantified majority extended lifespans dramatically. Disease nearly vanished. Cognitive decline delayed. The Guild remained small yet persistent. Mara aged naturally, her body bearing marks absent in optimized peers. Yet her music spread quietly through underground channels. Listeners described experiencing emotions unmediated by algorithmic curation—tempos irregular, harmonies unresolved, durations unpredicted. Her work reminded audiences of sensations preceding measurement. In her late sixties Mara's health faltered. Friends urged implant adoption for therapeutic optimization. She declined again, not from rejection of medicine but fidelity to chosen life architecture. Caregivers from both quantified and unmeasured communities supported her. Luca visited often, bringing insights from longevity science without imposing monitoring. Their conversations softened earlier tensions. He admitted envy of her experiential richness. She acknowledged admiration for his sustained vitality. Neither path invalidated the other; each revealed trade-offs inherent in surveillance and autonomy.

Mara composed until her final months. Her last piece, The Uncounted Hours, premiered in a hybrid venue bridging optimized and unoptimized audiences. Musicians performed without tempo guidance systems. Listeners removed biometric interfaces for the duration. For ninety minutes the hall existed outside measurement. The music unfolded unpredictably—pauses stretching beyond expectation, crescendos emerging from silence. Some found it disorienting. Others wept.

After her passing, longevity ethicists cited Mara's life in debates on surveillance proportionality. She had lived fewer years than optimized cohorts but demonstrated an alternative flourishing grounded in autonomy over metrics. Policymakers introduced autonomy sanctuaries within monitoring frameworks—spaces and professions exempt from continuous data requirements. Participation in longevity systems remained dominant, yet the right to unmeasured living gained recognition as ethically valid.

Luca attended a memorial gathering in the Guild retreat where Mara had composed her later works. No sensors, no projections, only wind through trees and voices sharing memory. He realized that surveillance had extended his years but Mara had expanded their texture. Longevity, he understood, encompassed not only duration but the freedom to inhabit time without constant observation.

As twilight settled, participants listened to a recording of The Uncounted Hours beneath open sky. The music moved without metric anchor, dissolving boundaries between moments. Luca felt something unfamiliar: existence unquantified, yet complete. He carried that sensation back into the optimized world, where he continued designing longevity systems—now shaped by awareness that extended life must preserve spaces beyond measurement.

The future would remain monitored, healthier, longer-lived. Yet somewhere within it, thanks to lives like Mara's, would endure unmeasured hours where autonomy breathed freely.

Chapter 4

Algorithmic Authority and the Definition of a Good Life

As artificial intelligence becomes increasingly capable of predicting and optimizing human health trajectories, it acquires a form of authority that extends beyond medicine into the philosophical domain of how life itself should be lived. Longevity AI does not merely recommend treatments; it implicitly defines what constitutes optimal behavior, desirable physiology, and successful aging. These definitions arise from statistical models trained to maximize measurable outcomes such as lifespan, disease-free years, cognitive preservation, or metabolic efficiency. Yet behind every optimization target lies a value judgment. The ethical question is not whether AI can identify behaviors correlated with longer life—it clearly can—but whether algorithmic optimization should determine what counts as a good life.

Algorithmic authority emerges gradually through performance superiority. When AI systems outperform human clinicians in predicting disease risk or identifying effective interventions, institutional trust naturally shifts toward algorithmic recommendations. Clinical guidelines begin incorporating model outputs. Insurance reimbursement aligns with

AI-derived protocols. Individuals internalize algorithmic advice as objective truth rather than probabilistic suggestion. Over time, recommendations evolve into norms. Activities associated with longevity gain cultural prestige, while those correlated with risk acquire stigma. Authority consolidates not through coercion but through perceived accuracy. This transition raises ethical concerns when statistical optimization becomes indistinguishable from moral prescription.

Central to this issue is the distinction between health maximization and life meaning. Longevity models optimize variables measurable across populations, yet meaningful human life encompasses dimensions resistant to quantification: creativity, relational depth, exploration, sacrifice, joy, and purpose. Many of these dimensions involve behaviors not optimized for biological efficiency. Artistic immersion may disrupt sleep; caregiving may increase stress; adventure may involve physical risk; devotion to ideals may entail sacrifice. If algorithmic authority frames such behaviors primarily through health cost, it risks narrowing acceptable life paths to those aligned with longevity efficiency rather than existential fulfillment.

The ethical tension intensifies when algorithmic recommendations carry predictive framing. A system indicating that a behavior statistically reduces lifespan by measurable years exerts psychological influence beyond traditional health advice. Individuals may interpret deviation as irrational or irresponsible even when the behavior contributes profoundly to meaning or identity. Over time, social narratives may equate virtue with adherence to longevity optimization and vice with deviation. This moralization of health efficiency represents a subtle yet powerful transformation: life choices evaluated primarily by projected biological outcome rather than subjective significance.

Algorithmic authority also interacts with institutional structures. Employers, insurers, and public health systems increasingly rely on AI longevity models to design policies. Workplace wellness programs align with optimization metrics. Insurance premiums reflect adherence to predictive health protocols. Public campaigns promote algorithmically validated behaviors as civic responsibility. While each initiative aims at population health improvement, collectively they may converge toward a standardized model of desirable living. Ethical risk arises when diversity of life priorities diminishes under systemic pressure to conform to algorithmic ideals of longevity.

Another dimension of authority concerns uncertainty representation. AI models operate probabilistically, yet their outputs often appear deterministic in user interfaces. Risk percentages, projected lifespan curves, and optimization scores convey precision that may exceed underlying certainty. When individuals perceive predictions as definitive, algorithmic authority strengthens disproportionately. Ethical design must therefore ensure transparent communication of uncertainty, emphasizing that optimization targets represent statistical trends rather than prescriptions for individual flourishing. Without such clarity, algorithmic guidance may overshadow personal judgment unjustifiably.

Cultural plurality further complicates universal definitions of a good life. Societies vary widely in attitudes toward aging, risk, duty, and fulfillment. Some traditions prioritize collective responsibility over individual longevity; others emphasize spiritual development, artistic expression, or social harmony. AI longevity systems developed within specific cultural frameworks may embed implicit assumptions about desirable behavior patterns. Exporting such systems globally risks imposing homogenized life models under the guise of health optimization. Ethical longevity AI must remain adaptable to diverse conceptions of flourishing rather than enforcing singular norms.

Resistance to algorithmic authority is not inherently anti-scientific. Questioning optimization does not deny health benefits; it affirms that biological efficiency is not the sole measure of life quality. Ethical frameworks must preserve space for individuals to choose paths involving calculated risk or non-optimized behavior without stigma or penalty. The legitimacy of such choice rests on respect for autonomy and recognition that meaning and mortality are intertwined. A life extended but stripped of chosen priorities may not constitute genuine well-being.

Importantly, algorithmic authority can also be constructive when appropriately bounded. AI systems can illuminate consequences previously invisible, enabling informed decision-making. They can identify behaviors enhancing both health and meaning, such as social connection or purposeful activity. Authority becomes ethically acceptable when presented as guidance rather than prescription, supporting individuals' values rather than redefining them. The challenge lies in ensuring that optimization tools remain subordinate to human judgment rather than supplanting it.

Institutional governance plays a crucial role in maintaining this balance. Policies should prevent discrimination based solely on deviation from optimization protocols when such deviation does not harm others. Health systems should present recommendations alongside value-sensitive framing acknowledging multiple dimensions of well-being. Education should emphasize probabilistic reasoning and personal meaning alongside biological health literacy. These measures reinforce that longevity optimization is a resource for living well, not a template for living uniformly.

This chapter establishes a fourth foundational principle of AI longevity ethics: algorithms may inform but should not define the good life. The extension of lifespan is valuable only insofar as it supports lives chosen authentically by individuals and communities. AI's role is to expand possibilities, not narrow them into standardized patterns of optimized existence. Ethical longevity preserves diversity of aspiration even as it enhances biological capacity. The future of extended life must remain pluralistic, allowing many forms of flourishing to coexist within longer time.

The Life That Was Optimized

By 2060 the Global Longevity Index had become the most influential metric in human society. Individuals received dynamic scores integrating biological age, disease risk, cognitive resilience, metabolic stability, and behavioral optimization. The Index guided healthcare access, employment trajectories, and social recognition. High scorers embodied the ideal of extended vitality; low scorers signaled preventable decline. The system promised fairness through objective measurement. Few questioned its authority.

Jonas Hale ranked in the ninety-ninth percentile.

From adolescence his biology aligned effortlessly with optimization protocols. He slept precisely, metabolized efficiently, responded favorably to regenerative therapies, and maintained stable cognition under stress. His Index projections exceeded one hundred fifteen years. Institutions competed for his participation in long-horizon initiatives. He led the design team for Adaptive Cities, urban environments calibrated to maximize population longevity. His life represented the triumph of algorithmic living: healthful, efficient, statistically exemplary.

Yet Jonas carried a quiet dissonance he rarely acknowledged. His days unfolded according to optimization schedules: nutritionally balanced meals, cognitively stimulating work blocks, social interaction quotas, restorative intervals. Activities outside Index-positive parameters felt indulgent or irrational. He had declined opportunities involving irregular hours, physical risk, or emotional volatility. Relationships formed with similarly optimized partners whose trajectories aligned. Life flowed smoothly, predictably. He was admired, yet internally he sensed absence—an unquantified dimension missing from continuous efficiency.

The dissonance surfaced during a site evaluation in the coastal region designated for Adaptive City expansion. There he encountered a small enclave excluded from Index integration due to incomplete data compliance. Residents lived without continuous monitoring, engaging in fishing, craft, storytelling, and seasonal rituals. Their biological metrics likely trailed optimized populations, yet their communal vitality struck Jonas unexpectedly. Laughter carried across evenings. Children played without behavioral tracking. Elders told histories not framed by health trajectories. The enclave existed outside algorithmic authority.

Jonas returned repeatedly under the pretext of environmental assessment. He met Mira, a sculptor shaping driftwood into forms inspired by tides and erosion. Her life followed rhythms unaligned with optimization: long nocturnal work sessions, irregular meals, immersion in weather and material. Her Index score, if measured, would likely fall below societal norms. Yet her presence radiated a groundedness Jonas found disorienting. She spoke of time not as quantity but texture. "Years are not equal," she told him once, smoothing sand from carved wood. "Some expand. Some compress. The Index cannot see that."

Their conversations unsettled Jonas's certainty. He began noticing how optimization framed his perception. When Mira stayed awake sculpting through storms, his mind calculated cognitive fatigue risk. When she skipped meals, he envisioned metabolic variance. His responses arose automatically from Index logic. He realized that algorithmic authority had colonized his interpretive lens. Experiences were evaluated through predicted biological cost rather than intrinsic meaning. Efficiency had become his primary language.

One evening the enclave held a tidal festival marking seasonal currents. Fires lined the shore. Music followed shifting waves. Mira invited Jonas to participate. The event lasted through the night—sleep deprivation, irregular nutrition, emotional intensity—all variables flagged negative by Index standards. Jonas hesitated, then remained. Hours dissolved into movement and sound. At dawn he felt physical exhaustion yet unfamiliar elation. The experience defied optimization yet enriched something unmeasured. For the first time he sensed life beyond algorithmic framing.

Returning to Adaptive City operations, Jonas perceived subtle pressure underlying Index culture. Colleagues optimized leisure to maintain scores. Social gatherings followed health-positive parameters. Risk-taking diminished. Diversity of lifestyle narrowed. Efficiency prevailed, yet spontaneity faded. The system had extended lifespan but homogenized living patterns. Jonas confronted a realization: algorithmic authority had not coerced conformity; individuals had internalized optimization as virtue. He himself exemplified this internalization. He initiated a research inquiry into qualitative well-being among non-optimized communities. Preliminary findings suggested paradox: despite lower predicted lifespans, participants reported higher experiential richness and perceived life completeness. The data challenged prevailing assumptions equating longevity with flourishing. Institutional response was cautious. The Index remained foundational. Yet Jonas's status allowed exploration.

His relationship with Mira deepened. She resisted Index enrollment, valuing autonomy over optimization. Jonas oscillated between worlds—Adaptive Cities by day, enclave by intervals. He noticed physiological fluctuations during unoptimized periods yet also heightened creative thought and emotional resonance. The dichotomy exposed a false premise: that biological efficiency and meaningful experience must align perfectly. Human life, he discovered, thrived in tension between structure and freedom.

A turning point arrived when Adaptive City governance proposed mandatory Index integration for all surrounding regions, including the enclave. The policy aimed at equitable longevity extension. Residents would receive monitoring implants and optimization guidance. Many welcomed improved healthcare. Mira hesitated. Integration risked dissolving unmeasured cultural practices into standardized schedules. Jonas found himself drafting impact assessments for the policy he now questioned.

He spent nights along the shore, wrestling with responsibility. Algorithmic authority promised longer lives for enclave residents. Yet it might also redefine their existence around metrics foreign to their values. The ethical dilemma crystallized: should longevity optimization override diverse conceptions of a good life? Jonas recognized his role in shaping that outcome.

During the final policy hearing he presented findings beyond expected scope. He demonstrated that algorithmic living improved measurable health but narrowed behavioral diversity and experiential variance. He argued that longevity systems should expand options rather than prescribe norms. Integration should remain voluntary, preserving enclaves of unoptimized living as legitimate cultural forms. His stance surprised colleagues and endangered his Index-aligned reputation.

Debate intensified globally. Media framed conflict between extension and authenticity. Ultimately governance adopted hybrid policy: monitoring access offered universally but participation voluntary; optimization incentives removed; cultural autonomy zones protected. The decision acknowledged pluralistic good life definitions within extended lifespan society. Jonas's Index score dipped as he reduced adherence to strict protocols. He accepted the decline without alarm. Time spent in Mira's world expanded his sense of living. Their partnership embodied negotiated balance: she adopted selective medical advances; he embraced irregular rhythms. Neither abandoned longevity nor autonomy entirely. They co-created life textured by both.

Years later Jonas walked the same shoreline now designated Cultural Autonomy Preserve. Adaptive Cities glowed inland; the enclave persisted beside tides. Mira sculpted nearby, hair silvered, hands steady. Jonas's lifespan projection remained high though lower than youthful forecasts. The change felt irrelevant. He had discovered that years optimized without chosen meaning were incomplete, while years shaped by values—even at biological cost—felt full.

Watching waves erode and renew the shore, Jonas understood the lesson reshaping longevity ethics: algorithms may extend life, but humans define why it matters. The good life could not be computed. It had to be lived.

Chapter 5

Bias, Inequality, and the Longevity Divide

Artificial intelligence systems inherit the structure of the data on which they are trained. In longevity science, this inheritance carries profound ethical consequences because lifespan prediction and health optimization models rely on historical medical, environmental, genetic, and behavioral datasets. These datasets reflect real-world inequalities in healthcare access, socioeconomic conditions, geography, and cultural representation. As AI-driven longevity systems identify patterns and allocate interventions, they risk embedding those inequalities into the future of human aging. The ethical challenge is not merely preventing discrimination but preventing the amplification of existing disparities through predictive health technologies.

Bias in longevity AI arises through multiple mechanisms. Data representation is central: populations with extensive medical monitoring and long-term health records generate high-resolution datasets, enabling accurate modeling of aging trajectories. Populations with limited healthcare access or fragmented records produce sparse datasets, increasing uncertainty in predictions. AI models optimized for statistical performance often perform best on well-represented groups, inadvertently disadvantaging others. When longevity systems prioritize interventions based on predictive certainty or responsiveness, individuals from underrepresented populations may receive delayed or reduced access to life-extending care. The resulting disparity is subtle yet cumulative: unequal model confidence translating into unequal longevity outcomes.

Socioeconomic bias compounds this effect. Wealthier populations adopt health technologies earlier, producing richer behavioral and physiological data streams. Continuous monitoring devices, genomic sequencing, and preventive interventions create detailed health trajectories that refine AI models. Lower-income populations, lacking such technologies, remain underrepresented in training datasets. As predictive accuracy improves primarily for high-resource users, longevity optimization becomes progressively tailored to their biological patterns and lifestyles. Ethical risk emerges when AI systems designed to extend human life effectively optimize for the already advantaged, widening lifespan gradients across society.

Geographic disparities introduce further complexity. Longevity determinants vary across environments—nutrition availability, pollution exposure, infectious disease prevalence, climate stressors, and healthcare infrastructure. AI models trained predominantly in specific regions may generalize poorly elsewhere. Interventions recommended based on one environmental context may be ineffective or inappropriate in another. When global longevity systems deploy standardized optimization protocols without sufficient local adaptation, populations outside primary data regions experience reduced benefit. Ethical deployment requires continuous regional dataset diversification and contextual calibration to prevent global longevity inequality.

Genetic diversity presents another dimension of bias. Human populations exhibit substantial variation in genomic patterns influencing disease susceptibility, metabolic function, and response to intervention. Longevity AI integrating genomic data must account for this diversity. Yet genomic databases historically overrepresent certain ancestries, particularly those from regions with advanced biomedical research infrastructure. Predictive models trained on such datasets may misestimate risk or treatment response for underrepresented genetic backgrounds. This misestimation affects both clinical decisions and preventive guidance, potentially altering lifespan trajectories through inaccurate prediction. Ethical longevity AI must therefore prioritize inclusive genomic representation and ongoing validation across diverse populations. Cultural bias also influences longevity modeling. Health behaviors—diet, activity, social structure, caregiving practices—vary across cultures and shape aging patterns. AI systems interpreting behavior through dominant cultural norms may misclassify or undervalue alternative practices. For example, communal living arrangements, traditional diets, or nonstandard activity patterns may not align with optimization models derived from different societies. When such systems inform public health policy or individual guidance, they risk marginalizing culturally embedded health strategies. Ethical longevity frameworks must integrate anthropological understanding alongside biomedical data to respect pluralistic pathways to healthy aging.

Institutional bias further shapes longevity distribution. Healthcare systems, insurers, and governments may adopt AI tools reflecting organizational priorities. If predictive models indicate higher intervention success among certain populations, institutions may allocate resources preferentially to maximize measurable outcomes. While economically rational, this approach risks deprioritizing populations with complex health determinants or lower predicted responsiveness. Longevity ethics must challenge purely utilitarian allocation when it undermines equity. The goal is not equal outcomes—biologically impossible—but fair opportunity for lifespan extension across diverse circumstances.

Another mechanism of inequality arises through feedback loops. As longevity technologies benefit certain groups earlier, those groups accumulate extended healthspan, productivity, and socioeconomic stability. Their enhanced data further refines AI models in their favor. Meanwhile, underserved populations remain underrepresented and less optimized. Over generations, such feedback loops could produce stratified aging trajectories across societies. Longevity becomes not only unevenly distributed but structurally inherited. Ethical intervention requires proactive inclusion strategies before disparities compound irreversibly.

Measurement bias also warrants attention. Longevity AI often emphasizes biomarkers easily quantified—metabolic efficiency, inflammation levels, cardiovascular indicators. Yet health and aging encompass broader determinants including mental well-being, social cohesion, purpose, and resilience. These factors vary across cultures and socioeconomic contexts and may not correlate directly with standard biomarkers. If AI systems prioritize measurable physiological metrics while overlooking contextual determinants, they may undervalue populations whose health strengths manifest differently. Ethical modeling must therefore expand definitions of healthy aging beyond narrow biomedical parameters.

Mitigating bias in longevity AI requires multifaceted strategies. Dataset diversification is foundational: inclusive recruitment across socioeconomic, geographic, and genetic spectra. Model auditing must evaluate predictive accuracy across subpopulations continuously. Intervention allocation frameworks should incorporate equity weighting alongside effectiveness. Transparency in model limitations allows clinicians and individuals to contextualize predictions appropriately. Participatory design involving diverse communities ensures cultural alignment. These measures transform bias mitigation from technical adjustment into ethical commitment embedded in system design.

This chapter establishes a fifth foundational principle of AI longevity ethics: the extension of human life must not be stratified by inherited inequality. Longevity technologies should narrow health disparities rather than widen them. Ethical success lies in ensuring that AI-driven life extension reflects the diversity of humanity rather than privileging subsets of it. As societies enter an era where lifespan is increasingly shaped by technology, justice demands that the benefits of extended life remain broadly shared rather than selectively concentrated.

The Map of Unequal Years

The Longevity Atlas hung in every public health center across the world by 2058. It was a living projection, continuously updated, mapping average biological age acceleration across regions. Colors ranged from cool blue—slow aging—to deep red—accelerated decline. Governments used the Atlas to allocate resources, evaluate interventions, and guide policy. Citizens studied it with fascination and unease. It revealed a truth long sensed yet never visualized: the geography of time itself was unequal.

Amara Okoye first saw the Atlas at seventeen in the coastal clinic of her hometown, Lagos Meridian District. Her region glowed amber bordering red. The display explained gently: environmental stressors, nutritional variability, healthcare access gaps. Average projected lifespan—seventy-two. Across the ocean, Northern Arcadia regions shone blue: projected lifespan—one hundred three. The contrast struck her more than any statistic she had learned. Some places, it seemed, lived longer.

Her mother worked in community health outreach. Amara accompanied her through neighborhoods where aging signs appeared early—joint degeneration in forties, metabolic disorders in fifties, resilience stretched thin by urban strain. Yet she also saw vitality: communal caregiving networks, intergenerational households, adaptive diets born from necessity. Life persisted richly within constrained conditions. The Atlas, she sensed, captured biology but not fullness. Determined to understand the map, Amara pursued biomedical engineering through international scholarship. She studied in Arcadia's premier longevity institute, where environments were optimized: clean air corridors, nutrient-dense food systems, ubiquitous monitoring. Students rarely witnessed premature aging. Biological age stabilization was routine. The contrast with Meridian shaped her resolve. Longevity, she realized, followed infrastructure as much as genetics. The Atlas mapped inequality encoded into environment.

During research rotations she examined predictive models underlying Atlas projections. She discovered dataset imbalances starkly: Arcadian populations generated dense longitudinal data; Meridian datasets were sparse and fragmented. Model confidence intervals widened dramatically for regions like her home. Yet intervention allocation algorithms favored areas with highest predictive certainty to maximize measurable success. The logic perpetuated disparity: regions already advantaged received further optimization.

Amara presented these findings cautiously. Faculty acknowledged the imbalance yet emphasized practical constraints. Resource deployment required reliable prediction. Without accurate models, interventions risked inefficiency. She recognized the reasoning yet felt its ethical cost. Longevity was being extended preferentially where data was richest, not where need was greatest. The Atlas recorded inequality; the system reproduced it. Returning to Meridian after graduation, Amara joined local health networks. She mapped environmental stressors unmodeled in global datasets: heat island intensity, micronutrient variability in regional crops, psychosocial resilience patterns in extended families. She realized that longevity determinants in her region differed qualitatively from Arcadian contexts. Global models failed to capture these variables, underestimating local adaptive strengths while overestimating deficits.

She began building Meridian-specific datasets with community participation. Residents contributed biological samples, environmental readings, and lifestyle narratives. Unlike external monitoring programs, her initiative emphasized reciprocity: participants accessed resulting insights directly. Data ownership remained local. Over years the dataset grew dense enough to train regional longevity models. Results surprised global observers. When contextual factors were included, Meridian aging acceleration proved lower than Atlas projections had suggested. Communal caregiving and adaptive nutrition mitigated environmental stress significantly.

Amara presented revised projections at the Global Longevity Forum. The new map displayed softened gradients; Meridian shifted from red toward amber. Yet the deeper message transcended color change. She demonstrated that longevity inequality arose partly from model bias—global systems misinterpreting regions through incomplete data. Correcting representation altered perceived destiny.

Policy implications followed. Intervention allocation algorithms incorporated equity weighting favoring high-need, low-data regions. Infrastructure investment targeted environmental determinants rather than solely biomedical optimization. Collaborative modeling centers formed across underrepresented regions, diversifying global datasets. Over decades the Atlas colors gradually converged—not identical, but less extreme.

Amara aged alongside these changes. By her fifties Meridian projections approached ninety. Clinics integrated local models guiding nutrition, heat adaptation, and stress resilience rooted in cultural practices. Youth grew without internalizing red-coded destiny. The Atlas remained, yet its gradients softened into nuanced mosaics rather than stark divides.

One evening Amara visited the coastal clinic where she first saw the map. The projection shimmered above new interactive panels showing local data contributions. Children explored simulations predicting how environmental improvements shifted aging curves. She felt quiet astonishment at transformation achieved through representation. Longevity had not been imported; it had been recalibrated from within.

A girl approached her holding a tablet displaying Meridian's current projection: average lifespan ninety-two. "Is this real?" the child asked. Amara smiled. "It's becoming real," she replied. "Because we learned to measure ourselves."

Later she walked the shoreline where city lights met sea horizon. The Atlas had once shown her world aging faster than others. Now she knew the deeper truth: inequality had been mapped not only in biology but in attention. Regions unseen by data were assumed to age poorly. Visibility itself altered outcome. Longevity ethics, she realized, required seeing fully before predicting.

Years later global longevity leaders cited Meridian's transformation as paradigm shift. AI models expanded to integrate diverse environmental and cultural determinants. Atlas gradients continued narrowing. Not uniformity, but justice approached. Human lifespan remained variable, yet its variability reflected biology rather than neglect.

Standing at the water's edge beneath vast sky, Amara considered the map that had shaped her life. It still glowed somewhere in the world, colors shifting with data. Yet she no longer saw regions of unequal years. She saw humanity learning to measure life more fairly, and through that fairness, extending it.

Chapter 6

Consent, Control, and the Ownership of Life Data

Artificial intelligence–driven longevity systems depend on continuous accumulation and analysis of personal biological data across the lifespan. Genomic sequences, metabolic markers, microbiome composition, behavioral patterns, environmental exposures, cognitive performance, and physiological signals together form longitudinal profiles that enable predictive modeling of aging trajectories. These data streams are not episodic medical records but persistent representations of an individual's biological existence over time. As longevity technologies mature, a central ethical question emerges: who owns the data that defines a human life, and who has the authority to control its use in extending or shaping that life?

Traditional medical ethics treats health information as confidential patient data managed by healthcare providers. AI longevity systems disrupt this framework by integrating data from multiple domains beyond clinical settings. Wearables, home sensors, digital platforms, workplace monitoring, and genomic services contribute to unified longevity profiles often stored and processed by private entities or distributed networks. Individuals may generate data continuously without clear awareness of its aggregation or downstream applications. Consent provided at initial collection may not encompass future uses enabled by evolving analytic capability. Ethical governance must therefore reconsider whether existing consent models remain valid in contexts where data persists and compounds across decades.

The temporal dimension of longevity data complicates ownership. Biological information gathered in early life retains predictive relevance throughout adulthood and aging. A genomic sequence obtained in youth may inform risk assessments, insurance decisions, or therapeutic targeting decades later. Behavioral data accumulated over years reveals patterns affecting future health forecasts. Unlike discrete medical events, longevity data forms an evolving narrative of the body across time. Individuals may wish to revoke consent or restrict usage, yet data already integrated into predictive models cannot be fully withdrawn without undermining system integrity. Ethical design must address how consent can remain meaningful when data permanence exceeds individual control.

Control over longevity data also intersects with economic value. Aggregated biological datasets drive innovation in pharmaceuticals, preventive medicine, and health optimization technologies. Organizations leveraging such data may derive substantial financial benefit. Individuals, however, typically receive only indirect health insights in exchange for contribution. This asymmetry raises questions of fairness: should individuals share in the value generated from their biological information? Is longevity data a personal asset, a communal resource, or a commodity subject to market exchange? Ethical longevity systems must determine appropriate frameworks for benefit sharing and recognition of data contributors' role in advancing life-extending knowledge.

Privacy concerns extend beyond exposure of sensitive information. Longevity data reveals intimate aspects of physiology, behavior, and potential future health states. Unauthorized access could enable discrimination or stigmatization based on predicted aging trajectories. Even authorized use may alter power dynamics between individuals and institutions. Employers or insurers accessing longevity profiles could influence opportunities based on projected healthspan. Governments could shape policy using aggregated data affecting population subgroups differently. Ethical safeguards must therefore ensure that control over personal longevity information remains primarily with individuals, limiting institutional leverage derived from predictive insight.

Consent complexity increases when considering relational data. Genetic and environmental information inherently involves family and community context. One individual's genomic sequence reveals partial information about relatives. Shared environmental exposure data implicates cohabitants or neighbors. Participation in longevity monitoring thus extends beyond personal boundaries. Ethical frameworks must consider collective consent dimensions, balancing individual autonomy with relational implications. Policies governing data use must acknowledge that ownership of biological information is rarely purely individual.

Another dimension of control concerns algorithmic inference. Longevity AI can derive new information from existing data, predicting disease risk, behavioral tendencies, or lifespan probabilities not explicitly provided by individuals. These inferences may be more sensitive than original inputs. Individuals may consent to data collection without anticipating conclusions drawn from it. Ethical governance must determine whether inferred information falls under original consent or requires renewed authorization. Without such consideration, AI systems may generate actionable knowledge about individuals beyond their intended disclosure.

Data portability and accessibility represent further ethical considerations. Individuals contributing to longevity systems often lack full access to their own integrated profiles. Data stored across multiple platforms may be fragmented or technically inaccessible. Lack of transparency undermines autonomy and informed decision-making. Ethical design should ensure that individuals can view, interpret, and transfer their longevity data across systems. Ownership implies not only legal rights but practical usability. Without accessible control interfaces, ownership remains theoretical rather than operational.

Global variation in data governance complicates longevity ethics. Different jurisdictions maintain distinct privacy laws, consent standards, and commercial regulations. Longevity data may cross borders through cloud storage or multinational research collaborations. Individuals' control over their biological information may weaken when transferred to regions with differing protections. Ethical longevity frameworks must therefore address transnational governance, ensuring consistent respect for consent and ownership regardless of data location or institutional context.

Technological solutions offer partial mitigation. Decentralized data architectures, encrypted personal data vaults, and user-controlled access protocols can enhance ownership. Blockchain-based consent tracking may allow granular permission management across time. Differential privacy techniques can protect individual identity while enabling aggregate analysis. Yet technology alone cannot resolve ethical questions of value distribution and authority. Governance structures must define rights and responsibilities associated with life data explicitly, integrating legal, economic, and moral perspectives.

This chapter establishes a sixth foundational principle of AI longevity ethics: individuals must retain meaningful ownership and control over the data that defines their biological lives. Longevity technologies should empower rather than appropriate personal information. Consent must be continuous, transparent, and revocable to the extent technically feasible. Benefit sharing should recognize contributors' role in advancing lifespan extension. The future of AI longevity depends not only on what data can reveal about life, but on who decides how that revelation is used. Ethical longevity respects that the story of a human body belongs first to the person living it.

The Archive of Her Life

In 2065 every citizen possessed a Life Archive: a comprehensive digital record integrating biological, behavioral, environmental, and experiential data from birth onward. The Archive enabled predictive longevity guidance tailored with extraordinary precision. Medical crises had nearly vanished. Disease onset was forecast years ahead. Lifespans lengthened steadily. Participation was considered both privilege and responsibility. Most individuals rarely considered who controlled the Archive; its benefits felt self-evident.

Leena Das had lived her entire fifty-two years within the system.

Her Archive began with neonatal genomic sequencing and expanded continuously through implants, wearables, cognitive assessments, and environmental sensors. The interface displayed evolving projections: biological age stabilized at forty-one, lifespan horizon one hundred eight. She worked as a historical archivist in the Cultural Memory Institute, curating analog artifacts from pre-digital eras. Ironically, she never questioned her own data archive's governance. It simply existed—trusted, immutable.

The first doubt emerged during her father's illness. Despite advanced prediction systems, his health declined unpredictably after a rare neurological condition. Reviewing his Archive, clinicians identified anomalies but could not fully reconstruct the trajectory. Segments of data were missing—collected by early devices now obsolete, stored on platforms no longer maintained. Ownership transfers between companies had fragmented his life record. The Archive that promised continuity proved discontinuous. Her father's predictive model failed partly because his data history was incomplete and inaccessible.

Leena requested access to his full Archive to understand what had been lost. Permissions required institutional authorization and corporate mediation. Weeks passed. When fragments arrived, she discovered that portions had been licensed for pharmaceutical research decades earlier. Those segments were no longer retrievable individually; they existed only within aggregated datasets. Her father's biological past had dissolved into anonymized models beyond personal recovery. The realization unsettled her profoundly. The Archive she assumed permanent was conditional, shaped by agreements made long before she understood them.

Her concern deepened when she explored her own Archive permissions. Terms signed during adolescence allowed broad analytic use across health, research, and commercial development. Consent renewals occurred automatically through system updates she had rarely read. Her life data—every physiological fluctuation, cognitive pattern, environmental exposure—had contributed to models generating immense institutional value. She had benefited through personalized health guidance, yet control over usage lay largely outside her awareness. The Archive belonged partly to her and partly to networks she did not fully perceive.

Driven by professional instinct, Leena began mapping data ownership chains across longevity systems. She traced how individual contributions flowed into corporate research pools, public health databases, and predictive infrastructures. Ownership diluted with each integration. Individuals retained nominal rights yet lacked practical authority. Data permanence favored institutions. Withdrawal of consent proved technically infeasible once integrated into models. The Archive, she realized, was less a personal record than a distributed ecosystem of her biological existence.

Her investigation reached a turning point when a multinational longevity corporation announced a new intervention derived from decades of Life Archive data. Marketing emphasized collective contribution. Shareholders profited. Contributors received gratitude but no tangible return. Public discourse questioned fairness. Leena joined a coalition advocating Life Data Sovereignty: recognition of individuals as primary stakeholders in biological information ecosystems. The movement argued that longevity knowledge derived from human lives should share benefits with those lives.

Resistance was immediate. Institutions warned that fragmenting data ownership would hinder research. Economists argued that innovation required centralized datasets. Yet Leena's coalition proposed alternative architectures: personal data vaults granting controlled access to researchers; revenue-sharing models distributing profits from derivative therapies; dynamic consent interfaces allowing individuals to manage participation across domains. The concept reframed longevity data not as institutional asset but collaborative resource.

Public interest surged after investigative reports revealed disparities in Archive usage across socioeconomic groups. Wealthier contributors negotiated favorable data terms; marginalized populations provided disproportionate data with minimal oversight. The ethics of life ownership entered global debate. Governments convened panels; corporations revised policies cautiously; individuals examined permissions long ignored.

Leena testified before the International Longevity Council. She described her father's fragmented Archive and her own diffuse ownership. "Our bodies generate the knowledge extending human life," she said. "Yet we rarely control how that knowledge lives beyond us. Longevity should not require surrender of life narrative." Her words resonated widely.

Policy drafts emerged recognizing individuals as co-owners of derived data value.

Reform unfolded gradually. New systems allowed individuals to store primary longevity data in encrypted personal vaults. Access required explicit, revocable permission. Aggregated models referenced encrypted derivatives rather than raw data. Contributors received dividends when therapies developed from pooled information reached market. Archives regained personal continuity; institutions retained analytic capability. Ownership shifted from abstraction toward shared governance.

Years later Leena visited the Cultural Memory Institute exhibit titled The Body Remembered. Displays traced evolution from early medical records to Life Archives and data sovereignty frameworks. She contributed her own Archive under new consent terms: accessible for research aligned with public health equity, restricted from purely commercial exploitation. For the first time she felt her biological story remained authentically hers even as it informed collective longevity knowledge.

At home she accessed her Archive interface. Visualizations spanned decades—growth curves, metabolic rhythms, cognitive patterns, environmental maps. Yet alongside metrics now appeared ownership indicators: permissions granted, benefits returned, collaborative projects supported. The Archive had transformed from silent observer into negotiated partnership between self and society.

She thought of her father, whose data had vanished into early systems without continuity or recognition. His life still informed models extending others' years, yet he had never known that legacy. Leena's generation would differ. Their data stories would remain visible and valued. Longevity knowledge would grow without erasing its human origins.

On a quiet evening she walked along the river near her apartment, sensors recording subtle physiological responses to dusk light. She smiled at the awareness yet felt no intrusion. The data flowed into her vault, under her terms, contributing to shared understanding while honoring individual ownership. She realized then that ethical longevity required not only extending life but respecting that each life generated knowledge uniquely its own.

Above the water, city lights reflected in shifting patterns—like data points across time, connected yet belonging to distinct sources. Leena watched them merge and separate with the current. The Archive of her life would continue expanding, yet its meaning remained anchored in her stewardship. Longevity, she understood, was not merely the accumulation of years. It was the preservation of authorship over the story those years created.

Chapter 7

The Risk of Redefining Aging as Failure

Artificial intelligence–driven longevity science increasingly frames aging as a modifiable biological process rather than an inevitable life stage. Predictive models identify molecular pathways associated with cellular decline, metabolic dysregulation, cognitive change, and systemic degeneration. Interventions target these pathways to slow, reverse, or optimize aging trajectories. This reframing has profound therapeutic promise, enabling prevention of disease and preservation of function across decades. Yet it also introduces a subtle ethical risk: when aging becomes technically preventable or optimizable, it may be culturally redefined from a natural transition into a correctable failure.

Historically, aging has occupied a complex place in human meaning systems. It represents both biological change and social transformation, marking shifts in role, authority, and identity. Eldership carries cultural significance in many societies, associated with wisdom, memory, and continuity. Physical decline, while challenging, has been interpreted as part of life's cyclical structure rather than individual fault. AI longevity technologies disrupt this narrative by positioning biological aging as a series of measurable deviations from optimal function. Cellular senescence, inflammation, and metabolic slowdown become targets for correction rather than phases to integrate. Ethical tension emerges when natural variability in aging is interpreted through optimization frameworks as deficiency.

The language surrounding longevity science contributes to this reframing. Terms such as age reversal, damage repair, and rejuvenation imply restoration toward an ideal baseline. Biological age metrics quantify deviation from youthful physiological states. Individuals with accelerated aging markers are described as experiencing decline relative to optimal trajectories. While clinically informative, such framing can influence self-perception and social attitudes. Aging bodies may be viewed less as evolving forms and more as problems requiring intervention. Ethical evaluation must consider how technological narratives shape cultural meanings of later life.

The concept of responsibility intensifies under this paradigm. If aging pathways can be influenced through lifestyle optimization, pharmacological interventions, and regenerative therapies, individuals may be expected to manage their biological aging proactively. Failure to adopt recommended protocols could be interpreted as negligence toward one's healthspan. This expectation parallels existing narratives around preventable disease but extends them to aging itself. Ethical concern arises when individuals experiencing natural decline are implicitly judged for insufficient optimization, ignoring genetic variability, environmental constraints, and personal priorities.

Healthcare systems may inadvertently reinforce this perception. Preventive longevity programs often track adherence to interventions associated with slowed aging. Metrics such as biological age reduction or inflammation control serve as indicators of successful participation. Individuals whose markers worsen despite engagement may feel personal failure. Those unable to access or comply with interventions may experience stigma. Aging thus shifts from shared human experience to individual performance outcome. Ethical longevity frameworks must resist transforming natural variation into moralized evaluation.

The commercialization of anti-aging interventions further amplifies this dynamic. Markets promoting youth preservation and vitality enhancement often frame aging as undesirable deterioration. AI-driven personalization increases perceived attainability of youthful physiology across extended time. Advertising narratives emphasize control over aging processes, suggesting that decline results from inadequate optimization. Such messaging risks marginalizing acceptance-based perspectives on aging and reinforcing unrealistic expectations of perpetual vitality. Ethical governance should address how commercial longevity narratives influence societal attitudes toward older individuals.

Another dimension concerns intergenerational perception. As extended vitality becomes more common among optimized populations, individuals experiencing visible aging may appear anomalous. Social comparison intensifies when biological age diverges significantly across peers. Older adults unable or unwilling to pursue interventions may face subtle exclusion from domains valuing sustained youthfulness. The ethical challenge lies in preserving respect and inclusion for diverse aging trajectories even as technological capacity to modify them expands.

Importantly, redefining aging as failure can erode psychological well-being. Acceptance of bodily change has historically supported resilience and identity continuity in later life. If decline is framed primarily as preventable, individuals may experience shame or regret when deterioration occurs. Such emotional burden compounds physical challenges. Ethical longevity must therefore balance intervention with acceptance, recognizing that even optimized aging remains finite and variable. The goal should be alleviating suffering rather than eradicating all signs of aging.

Cultural diversity again plays a critical role. Societies differ in attitudes toward aging, with some valuing maturity and others privileging youth. AI longevity systems developed in youth-centric cultures may implicitly embed negative assumptions about aging. Global dissemination of such frameworks risks homogenizing perceptions and diminishing traditional reverence for elderhood. Ethical integration requires respect for pluralistic aging narratives, ensuring technological capacity enhances rather than displaces cultural meaning.

From a philosophical perspective, aging contributes to life's narrative arc. Finite time shapes priorities, relationships, and generational continuity. Extended lifespan alters but does not eliminate this arc. Ethical concern arises when pursuit of perpetual youth obscures acceptance of temporal change. Longevity technologies should expand healthy years without denying the legitimacy of aging itself. Preserving dignity across all life stages remains essential to humane societies.

Practically, ethical longevity programs can counter failure narratives through inclusive messaging. Interventions should be framed as supportive rather than corrective, emphasizing enhancement of well-being rather than avoidance of decline. Biological age metrics should be contextualized within natural variability ranges. Healthcare communication should acknowledge uncontrollable factors influencing aging. Public discourse should celebrate diverse aging experiences alongside extended vitality. These measures sustain respect for individuals regardless of optimization status.

This chapter establishes a seventh foundational principle of AI longevity ethics: aging must not be pathologized as personal failure. Longevity technologies aim to reduce suffering and extend function, not to erase the legitimacy of growing older. Ethical societies recognize value in all stages of life, whether optimized or not. As AI enables unprecedented modulation of aging processes, humanity must preserve the understanding that dignity does not depend on perpetual youth. The success of longevity science should be measured not by elimination of aging, but by enhancement of life's fullness across time.

The Woman Who Was Supposed to Stay Young

In the city of Aurelia, aging had nearly disappeared from public view.

Advanced longevity protocols stabilized biological age across most of adulthood. Citizens in their seventies moved with the vitality once associated with middle age. Skin remained firm, cognition sharp, posture upright. Urban design, employment structures, and social imagery all reflected extended youthfulness. Aging still occurred eventually, but its visible signs were postponed so far into late life that earlier stages blurred into uniform vigor. Aurelia celebrated this achievement as humanity's liberation from decline.

Elisabet Marin had once embodied that triumph.

At forty she entered the Apex Longevity Program, achieving remarkable stabilization. For two decades her biological age hovered near thirty-five. She led corporate strategy initiatives requiring sustained cognitive endurance. Her appearance remained nearly unchanged. Friends admired her discipline in adhering to protocols—precise nutrition, regenerative cycles, metabolic calibration. She felt aligned with the city's ethos: aging was controllable, and she was controlling it.

The shift began subtly at sixty-three.

Routine markers indicated slight inflammatory drift and slower cellular repair response. Adjustments followed: intensified therapies, revised diet, increased monitoring. Yet the drift continued. By sixty-eight her biological age advanced to forty-eight. The number remained enviable by historical standards, yet within Aurelia's stabilized population it signaled deviation. Colleagues noticed fatigue in late meetings. Regenerative cycles required longer recovery. Elisabet perceived what others could not yet see: her body had resumed aging.

She responded with determination. Protocol adherence intensified; experimental interventions attempted. Gains proved temporary. Biological age crossed fifty-five. For the first time since Apex enrollment she confronted inevitability: optimization had limits. Her body was not failing; it was aging. Yet Aurelia's culture offered no narrative for graceful decline. Aging had become anomaly.

Workplace dynamics shifted. Projects requiring extreme endurance moved to younger-stabilized leaders. Invitations to high-intensity expeditions ceased. No discrimination occurred explicitly; optimization metrics guided assignments neutrally. Still, Elisabet sensed reclassification. She was transitioning from Apex exemplar to statistical outlier. The realization wounded more deeply than physical change. In a city that had deferred aging, to age felt like personal defeat.

Her self-perception eroded. She scrutinized mirror lines once invisible, posture once effortless. Metrics replaced meaning; each biological increment registered as loss. Friends reassured her, yet she detected subtle relief in their continued stabilization. Aurelia's success had rendered aging unfamiliar, almost suspect. Elisabet withdrew gradually from public roles.

One autumn morning she visited the Old Quarter, a district preserved from pre-longevity eras. There she encountered residents outside Apex programs—artists, historians, individuals who had declined stabilization for personal or philosophical reasons. They aged visibly yet moved with calm assurance. Wrinkles traced experience; hair silvered without concealment. Elisabet felt both curiosity and discomfort. These were the bodies Aurelia had transcended, yet they radiated dignity absent from her self-judgment.

She met Tomas, a retired sculptor in his eighties whose hands bore tremor yet shaped clay with deliberate grace. He spoke of aging as deepening rather than failure. "The body changes," he said, "but meaning accumulates." Elisabet struggled to reconcile this perspective with her internalized Apex ethos. Aging had been framed for decades as deviation from optimal function. Tomas treated it as evolution.

Visits to the Old Quarter became regular. Elisabet observed lives integrated with decline rather than resisting it: communal meals paced to slower rhythms, storytelling circles honoring memory, art celebrating imperfection. These practices contrasted sharply with Aurelia's relentless optimization culture. She recognized that Apex success had extended vitality yet narrowed acceptable aging narratives. When decline appeared, individuals lacked cultural language to hold it.

Her own markers continued advancing. Biological age reached sixty. She retired from corporate leadership, initially from necessity then by choice. Freed from performance metrics, she began volunteering in Old Quarter cultural programs bridging optimized and non-optimized populations. Younger-stabilized citizens visited reluctantly at first, encountering elders living fully despite visible aging. Dialogues emerged about limits of control and meaning beyond metrics.

Elisabet's transformation culminated when she agreed to speak publicly about her experience. The forum title unsettled her: Aging After Optimization. She described years of striving against inevitable change, the shame she had felt, and the liberation of encountering alternative aging narratives. "I believed I failed longevity," she said. "But longevity did not fail me. It gave me years. Aging gives them depth."

Her testimony resonated widely. Aurelia's institutions realized an unintended consequence of success: by delaying aging, they had pathologized its eventual return. Policies shifted. Public campaigns reframed aging as continuum rather than defeat. Intergenerational programs expanded. Apex protocols incorporated acceptance training alongside optimization. The city learned to integrate decline within extended vitality.

Elisabet aged into her seventies visibly now. Lines etched her face; gait slowed. Yet she carried new serenity. Freed from expectation of perpetual youth, she embraced phases once feared. Younger citizens sought her perspective not despite aging but because of it. She embodied transition Aurelia needed to acknowledge: longevity could extend youth but not abolish time.

One evening she stood on Aurelia's central terrace overlooking luminous skyline. Around her moved stabilized bodies glowing with health. She no longer envied or compared. Her own body told a different story—less efficient, more weathered, yet wholly hers. She felt continuity rather than loss. Aging had not expelled her from life; it had moved her deeper within it.

A young woman approached hesitantly, noticing Elisabet's silver hair and unoptimized features. "Does it frighten you?" she asked. Elisabet smiled gently. "It used to," she replied. "Now I know it's not failure. It's the next form of being alive."

As dusk settled over Aurelia, lights reflecting across glass towers, Elisabet sensed a quiet reconciliation spreading through the city. Longevity had extended life's surface; aging restored its depth. Together they formed a complete arc—years prolonged and meaning matured. And within that arc, no stage was failure.

Chapter 8

Dependency, Fragility, and the Risk of Technological Longevity

Artificial intelligence–driven longevity systems promise to extend human healthspan and lifespan through continuous monitoring, predictive modeling, and targeted intervention. As these systems integrate deeply into healthcare and daily living, they reshape the relationship between human biology and technological infrastructure. Longevity increasingly depends not only on intrinsic physiology but on external systems maintaining optimization. This dependence introduces a critical ethical consideration: when extended life relies on sustained technological support, human longevity itself becomes partially contingent on the stability, accessibility, and governance of complex technological ecosystems.

Historically, human lifespan improvements emerged from relatively robust interventions such as sanitation, vaccination, and basic medicine—technologies resilient to localized disruption. AI longevity, by contrast, relies on high-complexity networks: sensor arrays, data storage systems, computational models, supply chains for advanced therapeutics, and continuous clinical integration. Individual health trajectories may depend on uninterrupted monitoring and periodic intervention calibrated by algorithmic guidance. If such systems falter due to technical failure, economic collapse, geopolitical conflict, or infrastructure disruption, individuals whose physiology has adapted to optimized conditions may face rapid decline. Ethical longevity must therefore evaluate resilience alongside efficacy.

Dependency risk manifests biologically as well as infrastructurally. Interventions altering metabolic pathways, immune modulation, or cellular repair processes may create new equilibria reliant on continued modulation. For example, sustained suppression of inflammatory signaling or periodic regenerative therapy may maintain function that otherwise deteriorates. Discontinuation could produce rebound effects or accelerated aging. This phenomenon resembles dependency seen in certain chronic treatments, yet at systemic scale across lifespan modulation. Ethical assessment must consider whether long-term optimization creates physiological states less stable without ongoing technological input.

Economic fragility compounds this issue. Longevity technologies often involve proprietary platforms and specialized therapeutics requiring sustained financial support. Individuals benefiting from optimization may depend on continued access through insurance, employment, or personal wealth. Economic disruption could sever this access unevenly across populations. Those unable to maintain participation might experience disproportionate health decline relative to unoptimized baselines. Ethical frameworks must address whether life extension that cannot be reliably sustained constitutes responsible intervention or creates vulnerability tied to socioeconomic status.

Institutional continuity presents another dimension. Longevity AI systems evolve through iterative software updates, data recalibration, and clinical protocol refinement. Individuals' health management may rely on specific platforms over decades. If institutions providing these services dissolve, merge, or shift priorities, continuity of care could be compromised. Data incompatibility between systems may interrupt predictive guidance. Ethical design should prioritize interoperability and long-term stewardship, ensuring individuals are not bound to transient providers for survival-critical optimization.

There is also a philosophical dimension to technological dependency in longevity. Human identity historically encompassed adaptation to environmental variability and internal biological limits. AI longevity seeks to transcend these limits through engineered stability. While beneficial, such transcendence may reduce tolerance for natural fluctuation. Individuals accustomed to optimized physiological states may perceive normal variation as pathology. Societal expectations of sustained vitality may intensify intolerance for vulnerability. Ethical longevity must preserve psychological resilience and acceptance of biological change even within technologically supported healthspan extension.

Population-level fragility emerges when large cohorts depend on similar longevity infrastructures. Systemic failure—cyberattack, energy disruption, supply chain collapse—could impact millions simultaneously. Unlike distributed natural health determinants, centralized optimization systems create shared points of vulnerability. Ethical planning must incorporate redundancy, decentralization, and emergency protocols to mitigate such risk. Resilience should be considered integral to longevity technology design rather than ancillary.

Another consideration involves autonomy under dependency. Individuals reliant on continuous optimization may experience reduced capacity to disengage voluntarily. Choosing to discontinue intervention could entail significant health loss, constraining meaningful consent withdrawal. Ethical frameworks must acknowledge that sustained longevity modulation can create conditions where opting out becomes impractical. Transparency before initiation of such interventions is essential so individuals understand long-term implications of dependency.

Cultural implications also arise. Societies adopting technological longevity may shift values toward maintenance of optimized states. Natural aging or non-optimized living may be viewed as neglectful or inferior. This cultural bias can marginalize individuals who cannot or choose not to maintain technological support. Ethical longevity requires preserving pluralism in health trajectories, recognizing dignity in both technologically extended and naturally evolving lives.

Mitigating fragility involves multiple strategies. Designing interventions with reversible or adaptive profiles reduces biological dependency. Ensuring interoperability and open standards prevents institutional lock-in. Public investment in longevity infrastructure broadens access beyond market constraints. Emergency continuity planning safeguards optimized populations during disruption. Education fostering resilience and acceptance balances expectations of sustained vitality. These measures align longevity advancement with stability rather than vulnerability.

This chapter establishes an eighth foundational principle of AI longevity ethics: extended life should not create precarious dependence on fragile systems. Longevity technologies must be resilient, equitable, and sustainable across time. The success of life extension is measured not only by added years but by their security under varying conditions. Ethical longevity ensures that the pursuit of extended vitality strengthens rather than weakens human continuity in the face of uncertainty.

The City That Needed the System

Nova Helios rose from desert plains as the world's first fully optimized longevity city.

Every building integrated bioadaptive materials regulating air composition and microbial balance. Citizens wore seamless implants synchronizing physiology with environmental modulation. AI governance predicted disease trajectories across population networks, dispatching micro-interventions before symptoms arose. Average biological age remained thirty-five through decades. Nova Helios became symbol of humanity's triumph over aging variability—a city where life unfolded under continuous optimization.

Kai Ren arrived at twenty-six as infrastructure engineer.

His work maintained the central Longevity Core: the computational and therapeutic nexus orchestrating citywide health. The Core analyzed millions of data streams—metabolic patterns, immune markers, environmental shifts—calibrating interventions in real time. Citizens rarely perceived its operation. Vitality felt natural. Aging nearly absent. Kai admired the elegance: technology harmonizing biology across an urban ecosystem.

Years passed in seamless stability. Children grew without illness. Adults sustained youthful function. Elders remained vigorous. The desert outside city walls appeared harsher by contrast—communities aging normally under fluctuating conditions. Nova Helios stood as promise of optimized human future. Kai seldom questioned dependency implicit in that perfection.

The interruption came without warning.

A geomagnetic storm disrupted satellite arrays linking Nova Helios to external data and supply networks. The Longevity Core, designed with redundancy, maintained partial function initially. Yet predictive accuracy degraded as sensor synchronization faltered. Intervention schedules slipped. Environmental regulation lost fine calibration. For the first time in decades, physiological variability increased across citizens. Minor inflammation spikes, sleep disruption, metabolic drift—signals of unoptimized biology resurfacing.

Kai worked continuously to stabilize systems. Engineers restored connectivity gradually, but during the disruption he observed something unprecedented: citizens experiencing natural physiological fluctuation reacted with anxiety disproportionate to severity. Bodies accustomed to perfect regulation perceived ordinary variation as threat. Clinics filled with individuals distressed by symptoms once considered normal. Nova Helios had extended health yet reduced tolerance for imperfection.

More concerning were those requiring sustained modulation therapies. Some residents depended on periodic cellular repair cycles calibrated by the Core. Interruptions lengthened intervals beyond optimal thresholds. Aging markers advanced rapidly in these individuals compared to historically unoptimized populations. The city realized that its residents' biology had adapted to continuous support. Without it, decline accelerated.

The storm passed. Systems restored. Vitality returned. Yet collective awareness had shifted. Nova Helios had glimpsed fragility beneath its perfection. Longevity depended not only on biology but on uninterrupted technological orchestration. Kai could not forget the sight of optimized citizens confronting sudden vulnerability.

He proposed resilience redesign: decentralizing Longevity Core functions into distributed neighborhood nodes, integrating biological self-regulation training for citizens, and developing therapies maintaining stability without constant modulation. Leadership initially resisted. Nova Helios's identity rested on centralized precision. Decentralization implied imperfection. Yet public sentiment after the storm favored preparedness.

Kai led pilot programs in peripheral districts. Residents learned practices enhancing intrinsic physiological regulation—environmental exposure variation, adaptive nutrition cycles, psychological resilience training. Distributed nodes provided baseline support independent of central Core. Over years data revealed surprising outcomes: citizens with blended intrinsic and technological regulation maintained stability even during minor system fluctuations. Dependency lessened; resilience grew.

The transformation extended culturally. Public narratives shifted from flawless optimization toward partnership between biology and technology. Aging remained delayed, yet citizens accepted mild variability without panic. The desert beyond walls was no longer seen solely as deficiency but as reminder of intrinsic adaptability. Nova Helios matured from perfection toward robustness.

Kai aged into his sixties still youthful by global standards. He visited desert communities periodically, observing lives sustained without Core support. Their aging bore signs of time yet also resilience his city once lacked. He realized that longevity achieved through exclusive technological dependence risked brittleness, whereas blended systems fostered continuity.

A second geomagnetic storm struck decades later. This time Nova Helios endured smoothly. Distributed nodes maintained function. Citizens practiced adaptive behaviors instinctively. Health markers fluctuated slightly yet stabilized. No panic, no accelerated decline. The city had transformed dependency into partnership.

Standing atop a tower overlooking luminous districts and desert horizon, Kai reflected on the lesson learned. Technology had extended life magnificently, yet only when integrated with intrinsic human adaptability did it become secure. Longevity, he understood, required both system and self.

He addressed a gathering of young engineers entering Nova Helios's stewardship. "We once believed survival meant perfect control," he said. "But life endures through resilience. Our task is not to replace biology with systems, but to strengthen both together."

As dusk settled, city lights merging with stars above desert sky, Kai sensed harmony restored. Nova Helios still needed the system—but no longer depended on it blindly. Its people lived long not because technology erased fragility, but because they had learned to carry it wisely within extended years.

Chapter 9

Power, Governance, and Who Decides Longevity

As artificial intelligence becomes increasingly capable of predicting, modifying, and extending human lifespan, decisions about longevity shift from purely biological domains into spheres of power and governance. Longevity technologies do not operate in isolation; they exist within institutional, political, and economic systems that determine access, regulation, and direction of innovation. The ethical question is therefore not only how life can be extended, but who has the authority to decide how, for whom, and under what conditions longevity technologies are deployed. Control over lifespan extension represents one of the most consequential forms of power imaginable: influence over time itself.

Governance in longevity contexts encompasses multiple layers. Governments regulate medical interventions and public health policy; corporations develop and commercialize technologies; healthcare systems allocate access; research institutions shape knowledge trajectories; and individuals make personal decisions regarding participation. Each layer exercises influence over longevity outcomes. When these influences align with equitable and transparent principles, extended life can become broadly beneficial. When concentrated or opaque, longevity governance risks reinforcing inequality or coercion. Ethical analysis must therefore examine how authority over lifespan extension is distributed across institutions and populations.

One critical dimension concerns regulatory oversight. Longevity interventions often involve novel therapies, genetic modification, regenerative medicine, or continuous physiological modulation. Determining safety thresholds, approval criteria, and monitoring requirements requires regulatory judgment. Governments must balance innovation with protection against harm. Yet regulatory environments vary globally. Regions with permissive frameworks may accelerate development but expose populations to unproven interventions. Restrictive regimes may delay beneficial therapies. Ethical governance seeks harmonized standards ensuring safety without obstructing responsible progress. Transparency in decision-making fosters trust in longevity systems.

Corporate influence introduces additional complexity. Private entities often lead AI longevity research due to substantial capital requirements. Proprietary algorithms, data ownership, and intellectual property rights confer significant control over technology distribution. Profit motives may shape pricing, access tiers, and research priorities. While market incentives drive innovation, concentration of longevity capability within corporate domains raises ethical concerns about commodification of lifespan. Societies must determine acceptable limits of private authority over life-extending technologies and mechanisms ensuring public benefit from privately developed advances.

Public health governance intersects with individual autonomy. Governments may promote or mandate participation in longevity monitoring to reduce healthcare burden and enhance population vitality. Incentives such as insurance discounts or employment advantages may encourage adherence to optimization protocols. While beneficial at collective level, such policies risk coercing individuals into technological regimes shaping personal health trajectories. Ethical governance must balance population health goals with respect for voluntary participation and diverse conceptions of well-being. Longevity should remain opportunity rather than obligation.

Global disparities further complicate governance. Nations differ widely in resources, infrastructure, and demographic pressures. Wealthier regions may adopt advanced longevity technologies earlier, extending lifespan domestically while others lag. Migration patterns could shift as individuals seek access to life-extending care across borders. International governance frameworks must address equitable distribution and prevent longevity becoming geopolitical advantage. Collaboration in research and technology transfer can reduce global lifespan gaps, aligning extended life with shared human progress rather than national competition.

Another aspect involves democratic legitimacy. Decisions about longevity allocation—such as prioritization of interventions or public funding levels—carry profound societal implications. Should extended lifespan be guaranteed universally or allocated by market capacity? How should scarce therapies be distributed initially? Such questions transcend technical expertise and require societal deliberation. Inclusive governance mechanisms incorporating public input enhance legitimacy. Without participatory processes, longevity policy may reflect elite interests rather than collective values.

Ethical governance also requires accountability structures. AI longevity systems operate through complex algorithms potentially opaque even to developers. Decisions about risk classification or intervention eligibility may lack transparency. Individuals affected by such decisions need mechanisms for explanation, appeal, and correction. Oversight bodies must audit algorithms for bias and fairness. Governance should ensure that authority exercised by longevity systems remains answerable to those governed by it.

The intersection of longevity and social structure raises additional considerations. Extended lifespan influences retirement systems, workforce dynamics, and generational balance. Policies determining age-based entitlements or employment norms must adapt to longer healthspan. Governance decisions shape whether extended life enhances opportunity or entrenches generational inequality. Ethical longevity policy should distribute benefits across age groups and maintain intergenerational solidarity.

Finally, governance must anticipate long-term consequences of lifespan extension. Societal expectations, economic planning, and cultural narratives evolve as longevity increases. Decisions made during early adoption phases influence trajectories for generations. Concentrated power in shaping longevity systems could lock societies into inequitable patterns difficult to reverse. Ethical foresight demands that governance structures remain flexible, inclusive, and responsive to emerging impacts.

This chapter establishes a ninth foundational principle of AI longevity ethics: authority over life extension must be transparent, accountable, and broadly shared. Longevity technologies shape the temporal horizon of human existence; such power cannot reside exclusively in any single institution or interest group. Ethical governance ensures that extended life emerges from collective decision-making aligned with justice and dignity. The future length of human life should reflect shared values rather than concentrated control.

The Council of Years

In the late twenty-first century, lifespan extension had become technologically routine yet politically contested. Advanced regenerative therapies, genomic recalibration, and AI-guided aging modulation could extend healthy life beyond one hundred twenty years. Yet access varied widely across regions. Some nations guaranteed universal longevity care; others restricted it through market pricing. Migration pressures intensified as individuals sought jurisdictions offering extended lifespan. Conflicts arose over resource allocation, citizenship, and fairness. Humanity faced an unprecedented question: who should decide how long people live?

To address global tension, the World Assembly convened the Council of Years.

The Council brought together scientists, ethicists, policymakers, and citizens from diverse cultures to draft governance principles for longevity technologies. Its mandate extended beyond technical regulation to moral legitimacy: establishing how decisions about life extension should be made collectively. Observers doubted success; interests diverged sharply. Yet the stakes—authority over human time—compelled engagement.

Arin Solis arrived as youngest delegate, representing a cooperative longevity region where access was universal but modest. Her homeland prioritized equitable distribution over maximal extension. Average lifespan reached one hundred five, lower than affluent enclaves yet consistent across population. Arin carried quiet conviction that longevity should serve shared flourishing rather than competitive advantage.

Opposing perspectives filled the Council chamber. Representatives from high-extension zones argued that individuals should pursue maximal lifespan through market choice. Others from resource-limited nations demanded global redistribution of longevity technology. Corporate delegates defended intellectual property rights essential to innovation. Philosophers debated whether extended life constituted human right or optional enhancement. Consensus seemed distant.

Deliberations intensified when demographic projections revealed widening temporal inequality. Populations with access to advanced therapies could expect forty additional healthy years compared to those without. The divergence threatened geopolitical stability. Migration surges strained borders. Black-market longevity clinics proliferated. Governance vacuum amplified conflict. The Council confronted urgency: without shared authority, lifespan itself could become axis of global division.

Arin proposed reframing debate. “We speak as if years belong to nations or markets,” she said during plenary session. “But years belong first to persons. Governance must honor both individual aspiration and collective justice.” Her statement shifted tone. Delegates began exploring hybrid models balancing autonomy with equity.

Working groups drafted principles. Longevity access recognized as global human good rather than commodity alone. Core life-extending therapies designated universally accessible through international funding pools. Advanced elective extensions permitted under regulated markets. Technology transfer agreements ensured diffusion across regions. Algorithms governing allocation subject to public oversight. Decisions on new interventions required multilateral approval reflecting diverse populations.

Resistance persisted, especially from high-extension zones fearing dilution of advantage. Negotiations stretched months. Arin built alliances emphasizing mutual stability: equitable longevity reduced migration pressure and geopolitical tension. Economists demonstrated that broader healthspan extension increased global productivity benefiting all. Gradually compromise formed. No region would monopolize extended time; no individual denied baseline extension due to birthplace.

The Council ratified the Charter of Years. It established governance bodies overseeing global longevity distribution, research collaboration, and ethical review. Authority over life extension became shared among nations, institutions, and citizen panels. Implementation began unevenly yet steadily. Lifespan disparities narrowed over decades as therapies diffused globally.

Arin aged within this transformation. Her homeland's modest longevity rose as new therapies became accessible through Charter mechanisms. She traveled widely observing societies adapting to shared temporal horizon. Cultural exchange flourished among populations once divided by lifespan. Extended years became less marker of privilege and more common human condition.

In her eighties Arin returned to the Council chamber, now a living institution guiding longevity policy. Younger delegates debated emerging issues: cognitive preservation limits, reproductive timing across centuries-long lives, ecological balance under extended population presence. Governance evolved continually, reflecting shifting realities of human time. During a commemorative session marking Charter adoption anniversary, Arin addressed the assembly. "We once feared that whoever controlled longevity would control humanity," she said. "Instead we learned that sharing control preserves humanity." She looked across delegates from nations once unequal in lifespan, now converging in temporal expectation. Authority over years had transformed from competition into stewardship.

Later she walked along the river outside the Assembly complex. Lights from surrounding districts reflected across water—cities where people lived longer not through conquest but cooperation. She considered her own extended life, made possible partly by decisions forged in that chamber. Her years felt not owned by any institution but supported by shared governance honoring dignity.

A young observer approached, studying Council history. "Do you think humans should live even longer?" he asked. Arin smiled. "Perhaps," she replied. "But more important is how we decide together. Time is powerful. It must never belong to only a few."

As dusk deepened, the river carried reflections downstream toward distant seas. Arin sensed continuity: generations yet unborn would inherit frameworks ensuring that extended life remained collective achievement rather than concentrated power. The Council of Years had not solved all conflicts, but it had answered a foundational question. Who decides longevity? Humanity together.

Chapter 10

The Boundary Between Life Extension and Human Identity

Artificial intelligence–driven longevity technologies increasingly extend beyond preserving biological function into modifying the fundamental processes of human aging. Genetic editing, cellular reprogramming, neural preservation, and regenerative enhancement collectively shift the horizon of lifespan from decades toward potentially centuries. As these interventions deepen, the ethical question evolves from how long humans can live to what constitutes a human life across extended time. The boundary between life extension and transformation becomes blurred, raising profound concerns about identity, continuity, and the meaning of remaining human in an era of technologically mediated longevity.

Human identity has historically been anchored in biological continuity and finite lifespan. Developmental stages—childhood, adulthood, aging—structure personal narrative and social roles. Mortality imposes temporal limits shaping priorities, relationships, and generational renewal. Longevity technologies altering these temporal parameters inevitably reshape identity formation. When biological aging slows dramatically or reverses periodically, chronological age decouples from physiological state. Individuals may experience extended youth, delayed decline, or repeated regenerative cycles. Ethical reflection must consider whether such continuity preserves or transforms the self across prolonged duration.

Memory and psychological continuity are central to identity. Extended lifespan increases accumulation of experiences far beyond traditional human horizons. Cognitive preservation technologies may maintain neural integrity, yet the sheer volume of lived time raises questions about integration. Identity depends not only on memory retention but on narrative coherence—the sense that past, present, and anticipated future form a meaningful whole. Centuries-long life may challenge this coherence as contexts, cultures, and personal priorities shift repeatedly. Ethical longevity must examine whether technologies supporting extended memory also sustain integrative meaning rather than fragmenting selfhood across epochs.

The concept of self also intersects with bodily continuity. Regenerative interventions may replace or rejuvenate tissues repeatedly, altering physical substrates of identity. While cellular turnover already occurs naturally, accelerated or engineered renewal introduces qualitative change. If major physiological systems are replaced or extensively modified, questions arise about persistence of the same individual versus emergence of successive versions. Philosophical debates on identity persistence gain practical relevance when biological transformation becomes routine. Ethical frameworks should address thresholds of change beyond which life extension approaches transformation.

Social identity evolves alongside individual continuity. Extended lifespan alters generational relationships, cultural participation, and societal roles. Individuals living far beyond historical norms may witness multiple cultural eras, shifting communities and identities repeatedly. Belonging may become fluid rather than stable. Societies must adapt structures for education, vocation, partnership, and governance accommodating prolonged life trajectories. Ethical concern arises when identity stability erodes under constant adaptation to new contexts across centuries.

Longevity technologies may also enable elective identity modification. Neural enhancement, cognitive augmentation, or affective modulation integrated with life extension could reshape personality traits over time. Individuals might optimize memory capacity, emotional regulation, or cognitive style to maintain function across extended years. While beneficial, such modification complicates continuity. If personality evolves through technological intervention beyond natural development, determining persistence of original self becomes ambiguous. Ethical guidance should ensure voluntary and reflective engagement with identity-altering interventions.

Cultural and existential implications extend further. Human meaning systems have long been grounded in finitude: life stages, legacy, and generational succession. Extended lifespan challenges these frameworks. If individuals persist across multiple centuries, concepts of career, family lineage, and historical belonging transform. Societies may reconfigure narratives of purpose and achievement independent of traditional life arcs. Ethical reflection must explore how identity and meaning evolve when time itself expands beyond ancestral imagination.

Importantly, life extension need not negate humanity. Biological longevity enhancement can coexist with preserved identity if interventions maintain continuity of memory, agency, and narrative integration. The ethical boundary lies where enhancement transforms rather than supports these elements. Technologies preserving healthspan without altering core selfhood align with human continuity. Interventions reshaping cognitive or bodily substrate extensively may shift toward post-biological identity. Ethical discourse should distinguish between sustaining life and redefining species-level characteristics. Consent and autonomy again play critical roles. Individuals choosing deep longevity interventions should understand potential identity implications. Psychological counseling and longitudinal monitoring may help assess continuity over time. Societies should preserve rights for individuals to limit or discontinue transformation-oriented enhancements. The freedom to remain recognizably human must coexist with freedom to extend life technologically. Ethical longevity respects diverse thresholds of acceptable change.

Another dimension concerns intergenerational identity. Humans have historically defined themselves partly through relation to finite ancestors and descendants. Extreme lifespan extension compresses generational turnover, potentially altering cultural evolution. Individuals spanning multiple centuries may accumulate influence disproportionate to shorter-lived peers. Ethical governance must consider how extended identity interacts with societal renewal and diversity of perspective.

Ultimately, the boundary between life extension and identity transformation is not fixed but negotiated culturally and individually. AI longevity technologies amplify this negotiation by enabling interventions previously inconceivable. Ethical vigilance ensures that pursuit of extended years does not inadvertently dissolve the continuity that makes those years meaningful. Humanity's challenge is to extend life while preserving the narrative self that experiences it.

This chapter establishes a tenth foundational principle of AI longevity ethics: life extension must sustain human identity rather than erode it. Longevity technologies should support continuity of memory, agency, and meaning across expanded time. As the human lifespan stretches, ethical guidance must ensure that individuals remain themselves within extended years rather than becoming unrecognizable through transformation. The future of longevity depends not only on how long humans live, but on who they remain while living longer.

The Man Who Lived Too Long

Elias Varen had lived one hundred eighty-two years.

He was not unique in the late longevity era, yet his lifespan still evoked awe. Early adoption of regenerative cycles and neural preservation technologies had maintained his physiology near thirty-five for over a century. He had witnessed empires reorganize, languages shift, and continents reshape under climate restoration. His memory archive spanned epochs most citizens knew only from history modules. Elias embodied humanity's success in transcending biological limits.

Yet increasingly, he felt unmoored from himself. His earliest memories—childhood streets, parents' voices, youthful ambitions—remained vivid but distant, like stories read rather than lived. Centuries of experience layered over them: multiple careers, partnerships, identities. He had been scholar, explorer, policy architect, artist. Each phase genuine, yet collectively overwhelming. Narrative continuity blurred under accumulation. He struggled to answer a question once simple: who am I now?

Neural preservation ensured memory retention, but not integration. He could recall events with clarity yet sensed fragmentation between selves across decades. Values evolved repeatedly; goals abandoned; relationships outlived. The person who began life two centuries earlier seemed stranger rather than foundation. Elias realized that extending life had multiplied identity beyond intuitive coherence.

He sought guidance from Continuity Clinics, institutions specializing in psychological integration across extended lifespan. Counselors helped individuals map life phases into narrative arcs, emphasizing enduring themes. Elias charted centuries of choices. Patterns emerged—curiosity, creation, mentorship. Yet the exercise highlighted distance between early and current selves. He wondered whether continuity required transformation of memory itself rather than preservation alone.

During sessions he met Marael, a longevity ethicist researching identity thresholds. She proposed that human identity historically evolved within finite temporal span. Extending lifespan without evolving integrative mechanisms strained coherence. “We preserved memory,” she said, “but not necessarily selfhood.” Her work explored elective identity consolidation: cognitive processes allowing individuals to integrate vast experiences into cohesive narrative without erasing detail.

Elias resisted initially. His memories defined him; altering integration felt like losing authenticity. Yet fragmentation deepened. Encounters with individuals half his age revealed generational chasm. They admired his longevity yet could not relate to his layered past. He felt increasingly historical rather than present. Extended life risked detaching him from contemporary belonging.

He agreed to participate in Marael’s integration program. Neural interfaces guided reflective immersion across life phases, highlighting emotional continuities and dissolving obsolete identity constructs. Rather than deleting memory, the process recontextualized it within evolving self-concept. Early experiences regained immediacy as origins rather than relics. Later phases aligned along themes rather than chronology. Elias felt narrative coherence returning—not by compressing time but by weaving it. Integration altered his perception of aging. Though physiologically youthful, he acknowledged chronological depth without dissonance. Centuries no longer fragmented identity; they enriched it. He resumed mentorship roles, sharing perspective bridging eras. Younger citizens found resonance not in duration of his life but in continuity of his values. Elias recognized that longevity's success required psychological evolution parallel to biological extension.

Years later he addressed a gathering on extended lifespan identity. "I once believed living longer meant remaining unchanged," he said. "But identity must grow with time. Without integration, centuries scatter the self. With it, they deepen it." His words shaped emerging practices supporting narrative continuity across extended lives.

In private moments Elias revisited earliest memories now felt intimately present. His parents' voices carried through centuries not as echoes but as living threads. He realized he had not become different person across time; he had become expanded version. Longevity had stretched his story; integration restored its unity.

At two hundred he walked along coastal cliffs where he had played as child. Landscape transformed by restoration projects yet horizon unchanged. He sensed continuity across temporal vastness: same sea, same self experiencing it through ages. Extended life no longer felt alien to humanity. It remained human story lengthened.

A young companion asked whether he wished to live another century. Elias considered. Duration alone no longer defined value. “If I remain myself,” he replied gently, “then time is welcome.” The boundary between extension and identity had resolved: he had learned to remain human within extraordinary years.

As sun descended into endless ocean, Elias felt peace with life both long and continuous. Longevity had not erased his humanity. It had revealed its capacity to endure.

Epilogue

The Ethics That Must Outlive the Technology

Artificial intelligence has accelerated humanity into a new relationship with time. For most of history, lifespan was shaped primarily by biology and environment, with technology intervening only at the margins. Now, longevity itself is becoming an engineered outcome, influenced by predictive models, regenerative interventions, and continuous optimization systems. This transformation marks one of the most profound turning points in human history. Yet as this book has explored, the extension of life through AI is not merely a scientific achievement. It is an ethical frontier that will shape how humanity understands fairness, autonomy, identity, and meaning across generations yet to come.

The promise of AI-driven longevity is extraordinary. Diseases once considered inevitable may become preventable. Functional decline may be delayed or softened. Individuals may experience decades of additional vitality and cognitive clarity. Societies may retain accumulated wisdom and creativity longer than ever before. These possibilities carry genuine humanitarian potential. Reducing suffering and preserving capacity align deeply with the goals of medicine and human flourishing. Yet the same technologies capable of extending life also introduce risks that reach far beyond biology. They alter how societies distribute opportunity, how individuals perceive themselves, and how cultures interpret the value of aging and mortality.

A central theme throughout this book has been that longevity is never neutral. Decisions about life extension—who receives it, how it is delivered, and what trade-offs accompany it—reflect values embedded in technological design and institutional governance. Algorithms trained on unequal data can reinforce disparity. Systems optimized for efficiency may privilege certain populations over others. Continuous monitoring can enhance health while constraining autonomy. Commercial incentives can shape access to years of life itself. Without ethical vigilance, longevity technology may magnify existing inequities rather than alleviate them. Thus, the future of extended lifespan depends not only on scientific progress but on moral intention guiding that progress.

Another persistent insight is that extended life changes the meaning of being human. Aging has historically structured identity, culture, and generational continuity. AI longevity alters these structures by expanding temporal horizons. Individuals may live through multiple societal transformations within a single lifespan. Careers, relationships, and cultural belonging may become fluid rather than fixed. Memory accumulation across decades or centuries challenges narrative coherence. These shifts require psychological and cultural adaptation alongside biological intervention. Longevity technologies must therefore evolve with attention to identity preservation and meaning integration, not solely physiological function.

Equally important is the recognition that technological longevity introduces dependency. When healthspan relies on continuous monitoring and intervention systems, human survival becomes linked to infrastructure stability and governance integrity. Resilient design and equitable access are therefore ethical imperatives, not technical luxuries. Extended life should not be contingent on fragile networks or exclusive resources. The longevity future must be sustainable across economic, environmental, and geopolitical uncertainty. Ethical longevity strengthens human continuity rather than exposing it to new forms of vulnerability.

Throughout these chapters, a deeper question has emerged repeatedly: what is longevity for? Extending years is not inherently valuable unless those years support meaningful living. Health optimization must remain subordinate to human values rather than replacing them. Creativity, relationships, exploration, and purpose often involve behaviors that do not maximize biological efficiency. Ethical longevity frameworks must preserve space for diverse life paths rather than enforcing uniform optimization. The good life cannot be computed solely through biomarkers or predictive curves. It remains defined by human experience and choice.

The relationship between aging and dignity also demands careful preservation. As technologies slow or modify aging, societies may unconsciously redefine natural decline as failure. Yet aging remains integral to human narrative, marking transitions in role, perspective, and identity. Even in a longevity-enhanced world, variation in aging trajectories will persist. Ethical cultures must honor individuals across all stages of life rather than privileging perpetual youth. Extended vitality should expand possibility without erasing acceptance of change. Humanity's wisdom traditions, which integrate aging into meaning, remain relevant even as biology evolves.

Governance emerges as perhaps the most decisive factor shaping longevity's ethical trajectory. Authority over life extension carries immense power—power over time, opportunity, and generational balance. Concentrated control risks commodifying years or stratifying lifespan across populations. Shared governance, transparency, and accountability are therefore essential. Decisions about longevity must involve collective deliberation reflecting diverse perspectives. The extension of human life should be guided by humanity together, not determined by isolated institutions or markets alone.

At its deepest level, AI longevity confronts humanity with a paradox. Technology now offers increasing capacity to reshape biological limits, yet human fulfillment remains grounded in factors beyond technological control: connection, meaning, and identity continuity. The challenge is not choosing between natural aging and technological extension, but integrating them responsibly. Life can be extended without losing its narrative arc. Time can expand without dissolving selfhood. Aging can be modulated without erasing dignity. Achieving this integration requires ethical reflection equal in sophistication to technological innovation.

The future will likely witness further breakthroughs in regenerative medicine, genomic modulation, neural preservation, and systemic optimization. Lifespans may extend far beyond historical norms. Yet regardless of duration, the essential ethical questions will persist: Who benefits? Who decides? What remains human? The answers to these questions will determine whether longevity becomes a shared advancement or a divisive force. Technology alone cannot resolve them; they depend on values consciously chosen and maintained across generations. Ultimately, the ethical success of AI longevity will not be measured solely in added years but in how those years are lived and distributed. A just longevity future is one in which extended life supports flourishing for many rather than privilege for few. It is one in which individuals remain authors of their own narratives across expanded time. It is one in which aging, whether delayed or present, retains dignity and meaning. And it is one in which humanity's growing power over biology is matched by wisdom in its application.

As the reader closes this book, the technologies described here are already advancing. AI models are refining prediction of aging processes. Interventions targeting cellular decline are entering clinical practice. Monitoring systems are integrating into daily life. The ethical frontier is no longer theoretical. Each innovation invites society to choose how longevity will unfold. The responsibility belongs not only to scientists or policymakers but to all who live within extended time.

The future of longevity is therefore not predetermined. It will be shaped by choices made collectively about fairness, autonomy, identity, and meaning. Technology may lengthen life, but ethics must guide its direction. And in that guidance lies the possibility that humanity's greatest technological achievement—the extension of life itself—will also become its most humane.

www.ingramcontent.com/pod-product-compliance
Lightning Source LLC
LaVergne TN
LVHW022227200926
848481LV00006B/2158